Reliable knowledge

Canto is a new imprint offering a range of
titles, classic and more recent, across a
broad spectrum of subject areas and
interests. History, literature, biography,
archaeology, politics, religion, psychology,
philosophy and science are all represented
in Canto's specially selected list of titles,
which now offers some of the best and
most accessible of Cambridge publishing to
a wider readership.

Reliable knowledge

An exploration of the grounds for belief in science

JOHN ZIMAN, FRS

Emeritus Professor of Physics, University of Bristol

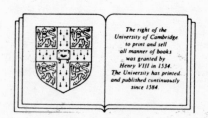

The right of the
University of Cambridge
to print and sell
all manner of books
was granted by
Henry VIII in 1534.
The University has printed
and published continuously
since 1584.

CAMBRIDGE UNIVERSITY PRESS
CAMBRIDGE
NEW YORK   PORT CHESTER
MELBOURNE   SYDNEY

Published by the Press Syndicate of the University of Cambridge
The Pitt Building, Trumpington Street, Cambridge CB2 1RP
40 West 20th Street, New York NY 10011-4211, USA
10 Stamford Road, Oakleigh, Melbourne 3166, Australia

© Cambridge University Press 1978

First published 1978
Reprinted 1979
Canto edition 1991

Printed in Great Britain by Billings and Son Ltd,
Worcester

British Library cataloguing in publication data
Ziman, John
Reliable knowledge: an exploration of the grounds for belief in science.
1. Science – Philosophical perspectives
I. Title
501

Library of Congress cataloguing in publication data
Ziman, John M 1925–
Reliable knowledge.
1. Science-Philosophy. I. Title.
Q175.Z55 501 78-3792

ISBN 0 521 22087 4 hardback
ISBN 0 521 40670 6 paperback

There is, upon the whole, nothing more important in life than to find out the right point of view from which things should be looked at and judged of, and then to keep to that point.

von Clausewitz

Contents

	Preface	*page*	ix
1	Grounds for an enquiry		1
	1.1	The challenge	1
	1.2	The theory	2
	1.3	The model	4
	1.4	Consensibility and consensuality	6
2	Unambiguous communication		11
	2.1	The language medium	11
	2.2	Mathematics as the ideal language	13
	2.3	Logical necessity	14
	2.4	The mathematical machine	17
	2.5	Metaphors and models	22
	2.6	The logic of experience	26
	2.7	Physics and physicalism	28
	2.8	Prediction	30
	2.9	The fit between theory and experiment	33
	2.10	Validating physics	38
3	Common observation		42
	3.1	Equivalent observers	42
	3.2	Pattern recognition	43
	3.3	Experiment	56
	3.4	Instrumentation	60
	3.5	Signal or noise?	64
	3.6	Discovery	70
4	World maps and pictures		77
	4.1	Material maps	77
	4.2	The map metaphor	82
	4.3	Pictures	85
	4.4	Paradigms	88
	4.5	Fallibility	92
5	The stuff of reality		95
	5.1	Perception	95
	5.2	'Artificial intelligence'	98

Contents

5.3 Extra-logicality 99
5.4 Intuition 101
5.5 Action and belief 105
5.6 Objectivity and doubt 107
5.7 The universality of science 109
5.8 Natural language 111
5.9 Cultural variations in cognition 116
5.10 How much is real? 119

6 The world of science 124
6.1 Specialization and authority 124
6.2 Learning science 126
6.3 Dissent and selection 130
6.4 Keeping in touch with reality 133
6.5 How *much* can be believed? 137
6.6 Parascientism 142
6.7 The limits of thought 148

7 Social knowledge 158
7.1 A science of behaviour? 158
7.2 Categorial imprecision 159
7.3 The algebra of social experience 163
7.4 Experimental simplification 166
7.5 Hidden variables 168
7.6 Models, toys and games 171
7.7 Simulations 174
7.8 Humanistic intersubjectivity 176
7.9 Origins of empathy 179
7.10 The limitations of a science of society 182

Index 187

Preface

This book has grown out of a lecture given in various places to a wide variety of audiences under the title 'Is Science to be Believed?' A generous invitation from the Van Leer Jerusalem Foundation encouraged me to reformulate, deepen and expand this into notes for four long seminars, which were presented in April 1975 to a helpfully critical group of philosophers, humanists and natural scientists. It seemed easy then to promise to 'write these notes up for publication'; but they had acquired a life force of their own, and it took another two years to break and tame them into the present text. Having no academic pretensions or professional affiliations outside physics, I have tended to read my own way into the diverse literature that is relevant to this enquiry, and to come to my own, perhaps idiosyncratic conclusions on many vexed issues of fact or principle. But I am grateful to Richard Gregory for some perceptive comments on the original notes, and to many others who have clarified tricky points by asking difficult questions in the lecture hall or in private conversation. Rosemary Fitzgerald gave valuable assistance in collecting the illustrations. And it is a pleasure once more, to express to Lilian Murphy my appreciation of her swift, accurate typing.

Bristol, June 1977

Note to the Canto edition

The 15 years since I began serious work on this subject have not changed significantly what I would want to say to the general reader. In spite of its informal style, this book was also intended to be, and remains, a serious challenge to other scholars in this field.

J.Z.

1
Grounds for an enquiry

> 'Science repudiates philosophy. In other words, it has never cared
> to justify its truth or explain its meaning.'
>
> Alfred North Whitehead

1.1 *The challenge*

This work arises from two sources: a *challenge* and a *theory*. The
challenge is to the beneficience of science as an agent of social change:
the theory concerns the nature of scientific knowledge.

The attack on science comes from many quarters, but is not well
concerted. The medley of opposition includes many strange
companions-in-arms, following contradictory causes. The conservative
fears that science will destroy the only world that he knows; the
progressive imagines that it will poison the paradise to come; the
democrat is cautious of the tyrannous capabilities of technique; the
aristocrat fears the levelling tendency of the machine. The pleas of
defence are equally inconsistent: some say that scientific progress is
automatic and inevitable; others that the future must be determined
by rational scientific planning; technocrats delight in telling us that
science will make life more comfortable; space addicts proclaim that
man must go forth and conquer the universe.

Science is such a complex human activity, so much a part of our
civilization, so rapidly changing in form and content, that it cannot
be judged in a few simple sentences.[1] We observe, nevertheless, that
some of the products of scientific technology have been damaging to
human welfare. In such cases, one can usually blame factors outside
the realm of science: too hasty innovation, subordination to unworthy
causes, distortion of social needs, or displacement of genuine human
goals. But the feeling has arisen that the evil factor is knowledge itself;
science is characterized as a materialistic, antihuman force, a
Frankenstein monster out of control.

More subtle critics[2] do not minimise the instrumental power of

---

[1] This statement needs no documentation. *The Force of Knowledge* (1976: Cambridge
University Press) is my personal view of the sociological and historical background
to the present work.
[2] Exemplified admirably by Theodore Roszak (1972) in *Where the Wasteland Ends*
(London: Faber).

science in its material, technical mode. The reliability of scientific knowledge in engineering, manufacturing, or medicine is not really in doubt. But they resist the attempt to extend science to the niceties of biological behaviour, human emotion and social organization. Any claim to scientific authority in such matters is regarded by such critics as pretentious, and inherently unsound. Other sources of insight and other guides to action must be treasured or sought beyond the reach of the scientific method.

This is the challenge – and it must be treated very seriously. A century ago, we might have described it as the conflict between Science and Religion. Nowadays, most people no longer base morality and aesthetics on divine revelation or rational theology; but no mature person with experience of life can seriously suppose that the issues of love and death, of justice and charity, could possibly be resolved by consulting the *Handbuch der Physik* or some latter-day edition of an *Encyclopaedia of the Behavioural Sciences*. On such matters, science clearly has little to say.

On the other hand it prejudges the issue to presume that a 'method' that has proved its worth in the realm of material technique can tell us nothing of value concerning man in society. We humans are part of the natural order of things, and subject to its necessities. The response to the challenge can be neither outright defiance nor abject surrender; the field of conflict is the middle ground, where the claims of science can be seen to be neither fanciful nor beyond reasonable doubt.

For this reason, the question of the *reliability* of scientific knowledge has become a serious intellectual issue. Once we have cast off the naive doctrine that all *science* is necessarily *true* and that all *true* knowledge is necessarily *scientific*, we realize that *epistemology* – the theory of 'the grounds of knowledge' – is not just an academic philosophical discipline. Very practically, in matters of life and death, our grounds for decision and action may eventually depend on understanding what science can tell us, and how far it is to be believed.

1.2 *The theory*

But what *is* science? How is it to be distinguished from other bodies of organized, rational discourse, such as religion, politics, law, or 'the humanities'? In an earlier work,[3] I have tried to show that scientific

[3] *Public Knowledge* (1967: Cambridge University Press).

2

knowledge is the product of a collective human enterprise to which scientists make individual contributions which are purified and extended by mutual criticism and intellectual cooperation. According to this theory *the goal of science is a consensus of rational opinion over the widest possible field.*

From this point of view, much can be understood about the ways that scientists are educated, choose research topics, communicate with one another, criticize and refine their findings, and relate to one another as members of a specialized social group. The 'consensus principle' thus leads directly into what is now called the *internal sociology* of the scientific community. From there we naturally proceed to investigate the place of science in society at large, trying to throw light on such important practical questions as the economics of research and development, the organisation of scientific institutions, priorities and planning of research, and the agonising ethical dilemmas facing the socially responsible scientist.

It is undoubtedly of great value to understand *how* science is done, and to appreciate the social role of the scientist and his institutions. But the epistemological challenge strikes deeper. What are the characteristic features of the body of knowledge acquired by this means? How does the consensus principle determine the *content* of science? What sorts of statement, about what aspects of the totality of things, are legitimate candidates for validation as 'public knowledge'? And to what extent does the striving for consensus eventually provide adequate grounds for belief and action?

In this book, therefore, I have deliberately turned away from the sociological aspects of science, to reconsider these fundamental intellectual issues. I am fully aware, of course, of the immense literature on the philosophy of science, where these very questions are asked again and again, and given a whole rainbow of answers. The writings of Plato and Aristotle, Bacon and Descartes, Kant and Wittgenstein, on this subject are the common heritage of our culture. But not being a trained philosopher, I could not pretend to be acquainted with all past and present opinions, all insights and all objections, on so large a topic.

Instead of attempting a general assessment of the epistemological problem, I propose to adopt the intellectual strategy of a typical paper in theoretical physics. A model is set up, its theoretical properties are deduced, and experimental phenomena are thereby explained, without detailed reference to, or criticism of, alternative hypotheses. Serious objections must be fairly answered; but the aim is to demon-

3

strate the potentialities of the theory, positively and creatively, 'as if it were true'.

The argument developed in the following pages is not, therefore, deeply embedded in the conventional philosophical literature. Ideas have been drawn from a variety of fields, such as linguistics, computer programming and anthropology, where I have had to sample and browse unsystematically on the look out for information or inspiration. In many particulars, however, the opinions expressed are far from novel, and are already to be found in the writings of some well-known philosophers.[4] I have done my best to cite such authorities – not only to do them justice, but also to bolster my own case. But I have not attempted to comb the literature for every hint of the same point of view – or for every possible objection to it; ignorance or neglect of work that might seem relevant to each particular issue is regrettable, but will, I trust, be forgiven.

1.3 The model

To characterize science fully, we must describe it in all its aspects – sociological, psychological and philosophical. For the purposes of this book, however, we need only consider a simplified model where the sociological dimension is reduced to a very simple 'Mertonian' scheme.[5] The relations between individual scientists (or between groups such as 'research teams') are assumed to approximate reasonably well to the Mertonian norms; in other words, they behave honestly to one another, both in communicating their own work and in accepting the work of others.

This idealization is essential if the epistemological issue is not to be hopelessly confused. We know, of course, that no scientific community is entirely healthy in this respect, and that the credibility of science as a whole is occasionally damaged severely by pathological deviations from the norms (6.5). Imperfections of communication or of critical analysis reduce the reliability of science in every field. In practice, however, this is seldom the dominant factor affecting credibility; the fragmentation and sectarianism characteristic of some fields of research (e.g. psychiatry) are not so much symptoms of social breakdown as consequences of the immense difficulty of making any progress at all in understanding the subject.

---

[4] For example, N. Capaldi (1975) in *Determinants and Controls of Scientific Development* edited by K. D. Knorr, H. Strasser and H. G. Zilian (Dordrecht: Reidel) presents a very schematic outline of the general point of view adopted in this book.

[5] See Robert K. Merton (1973) in *The Sociology of Science* (University of Chicago Press) pp. 267–78.

Although scientists often promise immeasurable future delights of understanding and truth, the epistemological challenge is always uttered at a particular moment: 'What can we believe *now*?' In assessing the credibility of scientific knowledge, we naturally look back over the past, but can put little weight on prognostications of an uncertain future. Our model, therefore, must be historically accurate, but need not be self-propelling; it will seldom be necessary to refer in detail to the psycho-dynamic forces that continually transform the contents of science (6.7).[6]

This is fortunate, since discussion of intellectual 'creativity' always tends towards a logical impasse – to cerebrate by other means than those of a particular science the unknown concepts that will eventually arise in that science. We shall see, indeed, that much more down-to-earth intellectual phenomena of belief and doubt, where the subject matter and context are known in advance to the psychological investigator, are also connected with 'creative' powers of imagination and intuition (5.4).

On the other hand, we cannot adopt a 'freeze-dried' model, where, on the appointed date, dispassionate, unprejudiced recording angels fly down to examine the scientific archives, and make an absolute assessment of the validity of each scrap of knowledge. As we saw in § 1.1, the epistemological challenge is not just an academic question; it arises in a human situation, and the answer is often required to deal with a human predicament. Those who ask the question, 'is this a matter on which science is to be believed?' must be given an answer that takes into account their own biographical experience and capabilities of comprehension. It would have been misleading, for example, to tell a railway engineer in 1920 that he should no longer believe in Newtonian mechanics because it had just been superseded by Einstein's general theory of relativity; for all his purposes, Newton's laws of motion remain as true as ever. From the very beginning I reject any system of *metascience* that purports to have such angels at its beck and call.[7]

---

[6] The numbers in parentheses are cross-references to other sections.

[7] This applies, in particular, to 'logical empiricism', in the various forms criticized by G. Radnitzky (1968) in *Anglo-Saxon Schools of Metascience* (Goteborg: Akademiforlaget). But I also, most emphatically, reject his hubristic view (p. xiv) that 'the metascientist will, one day, function like the business *consultant* – he will have to advise, warn, etc. in connection with the knowledge-producing enterprise, be it for the purpose of the planned production of some specific piece of knowledge or know-how, or be it for the regulation of the available "scientific capital" of a nation, a firm, etc. by means of foreign trade in scientific knowledge'.

1.4 *Consensibility and consensuality*

In its simplest form, therefore, our model consists of a number of independent scientists, linked by various means of *communication*. Each scientist makes observations, performs experiments, proposes hypotheses, carries out calculations, etc., whose results he communicates to his colleagues. As an individual, the scientist, like any other conscious being, acquires a great deal of personal knowledge about the world he inhabits, not only through his own experience but also through the information flowing to him from others. But when we talk of scientific knowledge, we refer to the content of the messages that accumulate and are available in the public domain, rather than to the memories and thoughts of each person.[8]

Going beyond this truism, we shall assume that scientific knowledge is distinguished from other intellectual artefacts of human society by the fact that its contents are *consensible*. By this I mean that each message should not be so obscure or ambiguous that the recipient is unable either to give it whole-hearted assent or to offer well-founded objections. The goal of science, moreover, is to achieve the maximum degree of *consensuality*. Ideally the general body of scientific knowledge should consist of facts and principles that are firmly established and accepted without serious doubt, by an overwhelming majority of competent, well-informed scientists. As we shall see, it is convenient to distinguish between a *consensible* message with the *potentiality* for eventually contributing to a consensus, and a *consensual* statement that has been fully tested and is universally agreed. We may say, indeed, that consensibility is a necessary condition for any scientific communication, whereas only a small proportion of the whole body of science is undeniably consensual at a given moment.

This model imposes constraints upon the *contents* of science. In the first place, fully consensible communication requires an unambiguous *language*, of which the ideal form is *mathematics* (2.2). But the exchange of logically consistent messages is fruitless unless they refer to recognizable and reproducible events within the experience of individual scientists; this explains the fundamental role of controlled observation and *experiment* (3.3) in the conventional 'method' of science.

But human cognition and communication are not restricted to pointer readings and algebraic formulae. Through our natural facility

---

[8] This is evidently 'World 3' of Karl Popper's *Objective Knowledge* (1972: Oxford University Press) – the logical contents of books, libraries, computer memories, etc. (5.5).

for *pattern recognition* (3.2) we may become aware of significant features of our experience, and transfer consensible messages, in the form of diagrams and pictures, whose 'meaning' cannot be deduced by formal mathematical or logical manipulation. For this reason, scientific knowledge is not so much 'objective' as 'intersubjective' (5.6), and can only be validated and translated into action by the intervention of human minds. In this respect, our model is less restrictive of the legitimate contents of science but offers less hope of strict testing of reliability than many conventional epistemological schemes.

These messages are not merely poured out into the archives nor passively received by other scientists. Consensuality implies strong interactions between the human actors in the drama. Thus, for example, elementary errors and misunderstandings are eliminated by the independent repetition of experiments, or by theoretical criticism. The fact that every competent scientist is trained – or moulded by bitter experience (3.4) – to the highest levels of self-critical precision in his communications (6.3) does not mean that this aspect of the model can be ignored; trivial errors are endemic in scientific research and must be continually corrected if the system is to generate anything approaching 'the truth'.

In the effort to maximize the area of consensus, however, the scientific community goes far beyond the exchange of easily corrected factual communications. Theoretical systems that explain the actual facts and imply a multitude of other potentially observable results are postulated. The consensuality of such systems is tested by such strategies as the attempted confirmation of predictions (2.8) or by the discovery of marginal phenomena that might prove inconsistent with accepted theories (3.6). It is important to realize that much of the research literature of science is intended *rhetorically* – to persuade other scientists of the validity of a new hypothesis or to shatter received opinions.

Looking now at our model as a whole, we easily recognize the power that participation in such activities has over the minds of individual scientists. Beyond the limits of their own personal observations of nature, they are aware of the immense body of results obtained by their predecessors and contemporaries, under stringent conditions of mutual criticism and reinforced by the persuasive authority of striking discoveries and astonishingly successful predictions. The relatively coherent and consistent set of beliefs thus generated is what we call a scientific *paradigm* or 'world picture' (4.4).

Nevertheless, despite its coherence and consensuality, such a para-

7

digm is not necessarily close to 'absolute truth'. As has been empha-sized, our model of science does not contain any independent source of 'objective' knowledge, and is therefore vulnerable to error in two significant ways.

In the first place almost every scientist is raised up, by formal education and research experience (6.2), within the world picture of his day, and cannot happily consent to statements that are obviously at variance with what he has learnt and come to love. The achievement of intersubjective agreement is seldom logically rigorous; there is a natural psychological tendency for each individual to go along with the crowd, and to cling to a previously successful paradigm in the face of contrary evidence. Scientific knowledge thus contains many fallacies (4.5) – mistaken beliefs that are held and maintained collectively, and which can only be dislodged by strongly persuasive events, such as unexpected discoveries or completely falsified predictions. In other words, our model must take into account the effects of its collective intellectual products on the cognitive powers of each of its individual members.

Secondly, and more significantly, is there any defence against the charge that the whole scientific paradigm is a self-sustained delusion (5.10)? The scientists in our model are almost always deliberately trained to a particular attitude to natural phenomena. How are their intellectual constructs to be distinguished from those of any other self-accrediting social group, such as a religious sect? What reason have we for preferring the scientific paradigm as the ideal, unique world picture?

We may assert that the social system of science is always open to the outsider (6.3), and that contributions of fact or opinion are not solely restricted to registered True Believers. It is well known, for example, that major scientific progress often comes from scientists who have crossed conventional disciplinary boundaries, and have no more auth-ority than a layman in an unfamiliar field. According to the ethics of 'the scientific attitude', science is valid in principle for Everyman, because *any* man could, if he wished, take up the study of science for himself, and would eventually be freely persuaded of its truth.

In practice, however, this is almost impossible; and when we look at the brainwashing implicit in the long process of becoming technically expert in any given branch of science, we see that it scarcely answers the objection – he who emerges from this process is no longer the unbiassed independent inspector who entered it ten years before.

More to the point, it must be emphasized that no scientist is a disembodied observing and conceptualizing instrument; he is a conscious human being, born and reared in the common life of his era. Long before he is taught about electrons, and genes, and exogamous fratries, he has acquired practical experience of pots and pans, cats and dogs, uncles and aunts. Although such mundane objects are seldom discussed as such in high science, they are not excluded from its realm. However fantastic it may appear on its wilder shores, the scientific consensus includes, by definition, the matter-of-fact, and must be coherent with everyday reality (5.10). Failure to accord with reliable 'commonsense' evidence is quite as discreditable as falsification of a theory by a contrived, abstruse experiment. Of course, commonsense evidence may often turn out to be irrelevant or ambiguous, but it cannot be trampled underfoot.[9]

The epistemological challenge to science thus leads to such profound questions as how each person acquires his view of the world, how far all men see the same world, and whether there can be any conceivable alternative to the 'reality' in which most men believe. The answers to these questions must not be anticipated here, for they determine the whole outcome of this book.

In some respects, however, this outcome cannot really be in doubt. Science does, after all, have its triumphs. It would be absurd to deny the validity of a theoretical system such as quantum mechanics, to which we owe our stocks of nuclear weapons. Who would doubt the credibility of Mendelian genetics, now completely confirmed at the molecular level by the deciphering of the genetic code? At least *some* of the knowledge that has been acquired 'scientifically' is as reliable as it could possibly be.

The basic strategy of this book is, therefore, to illustrate the workings of the social model of science by reference, initially, to the 'natural sciences', where the power of the 'scientific method' has been demonstrated beyond reasonable doubt. The most astonishing achievements of science, intellectually and practically, have been in *physics*, which many people take to be the ideal type of scientific knowledge. In fact, physics is a very special type of science, in which the subject matter is deliberately chosen so as to be amenable to quantitative analysis (2.7). But it is only when we have fully understood how

---

[9] In other words I accept the viewpoint summarized by G. Santayana (1962) in *Reason in Science* (New York: Collier Books) 'Science . . . is common knowledge extended and refined. Its validity is of the same order as that of ordinary perception, memory, and understanding. Its test is found like theirs, in actual imitation, which sometimes consists in perception and sometimes in intent. The flight of science is merely longer from perception to perception, and its deduction more accurate from meaning to meaning and from purpose to purpose.'

science really works even under the most favourable conditions that
we can appreciate its limitations. For that reason, I felt it necessary
to discuss the 'philosophy' of physics at some length, especially in
Chapters 2 and 3. Of course, this is difficult, because physics is a very
sophisticated intellectual discipline, whose techniques and attitudes
are not easily explained to the uninitiated; I hope that I have
managed at least to hint at some of this, by reference to various
historical and contemporary examples, without losing the reader on
the way. No doubt quite similar case histories could be found in
chemistry, geology, or biology, but they would not necessarily be any
easier to grasp out of context.

This investigation of the epistemology of the natural sciences takes
up the greater part of the book. It is only in the final chapter that
we arrive at a position from which we can begin to consider the
fundamental question of the book as a whole – how much ought we
to believe of what science might tell us about man as a conscious social
being, subject to unreasonable emotions and irrational institutions? I
do not pretend that such a question *can* be 'answered', but it seems
appropriate to subject it to a scrutiny based upon all that we have learnt
about the credibility of the natural sciences, where the subject matter
is so much easier to control. The results of this scrutiny are not, to
tell the truth, very favourable to the 'behavioural sciences' as we know
them today; perhaps, after all, the epistemological challenge is not
unjustified in that respect.

Needless to say, this inquiry is entirely concerned with the *cognitive*
aspects of science and not at all with any instrumental applications of
scientific knowledge to technology or other human activities. A successful application of knowledge is, of course, a pragmatic demonstration of its validity, and much of what is referred to as 'observation'
or 'experiment' in fact derives from carefully recorded practice.
Similarly, a confirmed or falsified prediction may have been derived
from a very practical event, such as the failure of a carefully designed
bridge. The main themes of this book may seem academic and aloof;
but in a society dazzled by silver-tongued technocrats and other self-
accrediting experts these questions are only a few breaths away from
harsh realities and bitter home truths.

2
Unambiguous communication

'Nature is written in mathematical language.'
                                          Galileo

2.1 *The language medium*

Since science is more than personal knowledge, it can consist only of
what can be communicated from person to person. The available
media of human communication determine the forms, and to some
extent the contents, of the messages that make up scientific knowledge.
To start with, as a crude 'zeroth-order approximation', we treat this
as a strict limitation; to achieve the ultimate goal of consensuality,
science must be capable of expression in an unambiguous public
language.[1]

Human beings communicate with one another in a multitude of
languages. In striving towards a consensus, we face the very ordinary
fact that some scientists publish their work in English, others in
Russian, others in Japanese. Not only does this lead to many practical
problems of mutual incomprehension, and substantial technical labour
in preparing glossaries, thesauri and classification schemes that span
several languages; it also limits scientific knowledge to what can be
translated unambiguously from one language to another.

This is not just a formal limitation that can be surmounted by
improved techniques of interpretation. It is well known, for example,
that poetry is essentially untranslatable. The message conveyed to a
Frenchman by, say, a poem by Verlaine cannot be rendered perfectly
to an Englishman in his own language. By their very nature, such
messages (alas) are so far from consensibility that they have little
prospect of contributing to the scientific consensus. This limitation is

---

[1] 'It is unreasonable to complain, as philosophers have so often done, because we *cannot
tell the truth without talking*' Stephen Toulmin: *The Philosophy of Science* (1953: London:
Hutchinson). Needless to say, beyond a certain stage of informality of presentation,
a scientific communication must be recorded in permanent form for reference in a
public archive. The invention of *writing* as a pre-requisite for the development of
science is emphasized by J. Goody (1977) in *Culture and its Creators* edited by J.
Ben-David and T. N. Clark (University of Chicago Press).

not avoided by the irresistible trend towards a single world-wide language for scientific communication. To be sure that his work will be read and understood, the Italian or Egyptian or Argentinian scientist translates it for himself from his native tongue into the international scientific language. This language is no longer Latin, but is, of course, English; or, rather, it is *Broken English*, for even those who speak and write it accurately and fluently seldom command it in all the depth and subtlety of a mother tongue. The consensible contents of such publications are thus no broader in scope than what can be accurately translated from one language to another by a competent scientific author.[2]

Science could not, in any case, use all the resources of a natural language such as English. Consensible communication demands the deliberate style or mood appropriate for the transmission of unambiguous knowledge – what Gilbert Ryle calls *didactic* discourse.[3] But consensibility is not enough; every scientist is pressing towards consensuality. His communications are intended not only to tell things as he saw them, or as he thinks they are; he is also desperately keen to *persuade* his readers or audience. A scientific message often has the purpose of changing a preconceived notion, demonstrating an unsuspected contradiction, or announcing an unexpected observation. It is addressed to an actual sceptic, a potential critic; it must be convincing, it must be watertight.

By a psychological inversion this rhetorical motive is best served by a very plain and modest style.[4] But in normal, natural language it is easy to slip out of the noose of a line of reasoning. Ordinary everyday verbal controversies are always loose and inconclusive; one side or the other finds loopholes, such as ill-defined terms or ambiguities of expression, that allow him to avoid an unpalatable conclusion. That is why legal documents have to be written in a complex, formalized (and ultimately repellent) language. Scientific communications are

[2] President de Gaulle once insisted that French scientists should speak French at international conferences. This was countered by the English participants who said that they would speak French too!

The very real difficulties of translating advanced scientific and technical information into rigidly traditional languages such as Arabic is discussed perceptively by C. F. Gallagher, 'Language rationalization and scientific progress' in K. H. Silvert (ed), *The Social Reality of Scientific Myth* (New York: American Universities Field Staff Inc.) pp. 58–90.

[3] 'It is talk in which, unlike most of the others, what we tell is intended to be kept in mind...to be remembered, imitated and rehearsed by the recipient...(It) can be accumulated, assembled, compared, sifted and criticized'. *The Concept of Mind* (1949: London: Hutchinson).

[4] See Ziman, *Public Knowledge*, p. 96.

forced along the same path. In the search for perfect precision and
overwhelming certitude, they become formalized statements in which
technical terms that have been previously defined with the maximum
rigour are bound together in unambiguous syntactical arrangements
implying complete logical necessity.[5] No wonder scientific writing lacks
literary grace and is denounced for its barbarity!

2.2   Mathematics as the ideal language

The ultimate step in formalizing a language is to transform it into
mathematics. As each word in the language becomes more and more
precisely defined, its specific meaning comes to reside in its relations
to other words; these relations acquire the force of axioms, akin to
those defining the essential relationships between say, the 'points' and
'lines' in Euclidean geometry. Two scientists who are familiar with
such a system of definitions and axioms can thus send each other
unambiguous messages. There is no danger of misunderstanding the
statement, 'the carbon atoms in benzene form a regular *hexagon*',
because a regular hexagon is a well-defined figure. The mathematical
concept of *number* is very precise. The statement that 'a neutral carbon
atom contains *six* electrons' is completely consensible, being comprehensible and capable of verification in principle, by any observer.

The ideal language for scientific communication is thus to be found
in mathematics. Of its essence, mathematics is unambiguous and
universally valid. Not only do modern Chinese and Indian mathematicians use the standard symbolism of European algebra; ancient
Chinese mathematicians discovered the 'theorem of Pythagoras' independently of their Greek contemporaries, and ancient Hindu mathematicians juggled with enormous numbers long before these were
needed in astronomical computations. The urge to express all scientific
knowledge in mathematical terms is an elementary consequence of our
model of science. In the pursuit of a consensus, we are bound to hit
upon this device for constructing messages with the maximum degree
of clarity and precision. Whatever we may eventually suspect concerning the limitations of a mathematical description of human experience, the central place of mathematics in the natural sciences is
well-deserved and appropriate.[6]

[5] E.g. 'The ultimate reason for formalization [of scientific theory] is that it provides
the best objective way we know to convince an opponent of a conceptual claim.'
P. Suppes (1968) *Journal of Philosophy*, **65**, 651.
[6] This implies that *mathematics is a social institution*, as pointed out by D. Bloor (1973)
Studies in the History and Philosophy of Science, **4**, 173.

Notice, however, that this way of polishing up the language of scientific communication does not make any particular message more *true*. The statement, 'a neutral carbon atom contains *seven* electrons' is precise, unambiguous, consensible, logically self-consistent, etc. – but it happens to be *false*. This seems a trivial remark, but a great deal of scientific nonsense is generated through failure to appreciate its force (5.3). It is all too easy to derive endless strings of interesting-looking but untrue or irrelevant formulae instead of checking the validity of the initial premises.

Mathematical language also has very limited descriptive powers. A Euclidean 'point', without size or shape, makes good sense as a dynamical 'particle', but is a poor sort of representation for a molecule or a planet. The process of formalization produces an abstract entity that satisfies its defining relations perfectly but which has been stripped of all other attributes. This is one of the fundamental objections to the use of mathematical techniques in the social sciences (7.3); the data, concepts and other entities that arise in the study of human behaviour are never so simple and unadorned as the objects and operations we have learnt to manipulate mathematically.

Thus, the use of mathematical language is a desirable, but not essential, characteristic of a branch of science. Natural language may be imperfectly consensible, but is infinitely richer in vocabulary than algebra. A sketched picture may convey far more than a geometrical definition. The first priority of science is that *meaningful* messages should pass between scientists, not that these messages should be censored to misleading triviality in the name of logical precision.

2.3   Logical necessity

Unfortunately, deficiencies of consensibility can be found even in the most sophisticated mathematical systems.[7] The study of the 'foundations of mathematics' reveals ambiguities and paradoxes which cannot be resolved by formal analysis. Although most of these deficiencies relate to the correct usage of concepts such as infinite sets of infinite sets, which seem of little practical significance, they cannot be ignored in any rigorous discussion of scientific epistemology. Gödel's theorem,[8] for example, asserts the existence of propositions that can

---

[7]   See e.g. *What is Philosophy* by Stephan Körner (1969: London: Allen Lane).
[8]   See e.g. 'Gödel's Proof' by E. Nagel and J. R. Newman in *The World of Mathematics* edited by J. R. Newman (1960: London: George Allen & Unwin) p. 1668. The essays by C. G. Hempel (pp. 1619–46) and by R. von Mises (pp. 1723–54) in this volume are admirable introductory accounts of the philosophy of mathematics.

be formulated, but whose truth or falsity cannot be decided, within a given axiomatic system. In other words, a scientific message communicated in the mathematical language generated by these axioms could contain fatally uncertain statements relying on such propositions.

In the analysis of such problems, we inevitably encounter a special branch of modern mathematics – formal *logic*. The fundamental importance of logic in science needs no emphasis.[9] At this point we see logical consistency or *logicality* as a necessary condition for meaningful intersubjective communication. A scientific message that was logically contradictory would be totally ambiguous and hence void. A patently illogical language would be quite useless as a medium for science. One of the advantages of a good mathematical symbolism is that it can avoid logical errors by intellectual automation.

But researches in the foundations of mathematics and mathematical logic have shown that the hope of finding a unique and *perfectly* logical language of this kind is vain. In the drive towards an absolute consensus, we eventually arrive at a point where differences of opinion concerning the status of the laws of logic itself could not be resolved by reference to a higher authority.

Logicality is not a *sufficient* condition for scientific discourse. It applies only to the *grammar* of the scientific language, and says nothing about the contents of the messages whose form it constrains. Consensual theorems of formal logic are an important branch of pure mathematics but are practically empty scientifically. Almost all science is based upon a variety of other principles that are shared by the community of scholars but are not deducible from logic alone. It is practically impossible, for example, to make a scientific statement that does not depend on such Kantian 'categories' as 'space' and 'substance'.[10] Much of the subject matter of pure mathematics has similar non-logical foundations.

Here we touch on one of the central issues of the epistemology of

[9] We might refer to a quotation from Max Weber 'In Greece, for the first time, appeared a handy means by which we could put the logical screws upon somebody, so that he could not come out without admitting either that he knew nothing, or that this and nothing else was the truth, the *eternal* truth that never would vanish as the doings of the blind men vanish.'

[10] P. F. Strawson (1966) in *The Bounds of Sense* (London: Methuen) p. 150 remarks (5.6): 'we should remember that all Kant's treatment of objectivity is managed under a considerable limitation, almost, it might be said a handicap. He nowhere depends upon, or even refers to, the factor on which Wittgenstein, for example, insists so strongly; the *social* character of our concepts, the links between thought and speech, speech and communication, communication and social communities...another name for the *objective* is the *public*.'

science. Philosophers have long been concerned with the investigation and characterization of the categorial frameworks[11] that must be shared if consensibility and ultimate consensuality are to be achieved. Scientists must already agree on a great many things if they are to come to agreement on something more. For the present, however, we are in no position to specify in advance, or to delimit hypothetically, the range of the 'supreme principles' in the cognitive sphere. We shall discover, indeed, that the practice of science, within the reality of human life, individually and collectively, develops and refines such principles (6.7). We cannot even be sure that they can be regarded as objective or *a priori*. A great deal of excellent scientific knowledge depends upon a widely shared human perceptual faculty – the mysterious skill that we call *pattern recognition* (Chapter 3). Yet this faculty does not seem amenable to complete logical analysis (5.3) and it is not perfectly uniform amongst all men. Going back to logic itself, we have no guarantee that the elementary forms to which we are so attached are absolutely universal; they may depend on the 'world-wide' characteristics of the only sentient beings with which we happen to be familiar – mankind with its vocalized languages and other cultural devices.[12]

Nevertheless, having been warned not to take formal mathematical reasoning entirely at its own estimation, we know that a scientific communication is almost valueless unless it is expressed in precise language and has a sound logical structure. These desirable qualities are most readily achieved by using mathematical concepts and symbolism. Quantitative measurement and mathematical theorizing do not automatically generate reliable scientific knowledge, nor are they essential for reputable research in every field of science; but in appropriate circumstances they contribute enormously both to consensibility and to consensuality. The remainder of this chapter will be concerned with the intellectual role of mathematics in science, and with its influence on the strategies of research and on the contents of our knowledge of nature.

[11] E.g. *Categorial Frameworks* by S. Körner (1970: Oxford: Blackwell).
[12] At least we can agree with D. Bloor in *Knowledge and Social Imagery* (1976: London: Routledge & Kegan Paul) p. 97: 'mathematics like morality is designed to meet the requirements of men, who hold a great deal in common in their physiology and in their physical environment'. The 'requirements' of course are those of unambiguous communication.

2.4   *The mathematical machine*

Mathematics is invaluable to science as a strong grammar for didactic discourse; it is the ideal vehicle for precise intersubjective communication. The clarity and universality of mathematical language is of the greatest practical importance. Scientific messages are not normally addressed directly from one particular scientist to another. An essential feature of our model of the 'scientific method' is the library or *archive* to which messages are communicated, and where they are stored for subsequent consultation (6.5). Science is *public* knowledge; it is the contents of this archive, and should not be extended indiscriminately to all that may be known or suspected about the world by all would-be scientists. A mathematically phrased statement in the archive is in the best form for consultation, comprehension, or critical assessment.

A further special advantage of mathematical messages is that they may be symbolized, manipulated and *transformed* according to precise rules, without loss of meaning. Suppose, for example, that we have observed 750 black swans and 250 white ones flying overhead. Our message could equally well have recorded that, 'of 1000 swans, 75% are black and 25% are white'. Or we could have said that 'the ratio of black to white swans is as three to one'. Or we could point out that 'there were 500 more black ones than white ones'. Or it might be noted that 'the probability of any one swan being black is 0.75'. Or we insist that 'if w is the number of white swans and b is the number of black swans, then $b+w = 1000$ and $b = 3w$'. Or we might even, for some esoteric theoretical reason, be pleased to report the remarkable fact that

$$\sin^{-1}\left(\frac{w}{b-w}\right) = \frac{\pi}{6}$$

– and so on, and so on. The rules of arithmetic, algebra, trigonometry, calculus, group theory, analytic function theory, etc. etc. permit us to generate an infinite variety of unambiguous statements, of varying degrees of complexity, all of which are logically equivalent to the original message.[13]

By academic tradition, mathematics straddles the boundary between the 'Arts' and the 'Sciences'. This ambivalence is justified. There is no doubt that a genuine mathematical conclusion must be completely consensual; a theorem, once satisfactorily proved, must be true every-

[13] H. Putnam (1975) in *Mathematics, Matter and Method* (Cambridge University Press) p. 43, makes this point.

where, at all times, for all men. But what is called *pure* mathematics is no ordinary science, and must be given a special place in our model (1.3). The pure mathematician is not concerned with the factual truth of any particular statement; his interest is in the logical status of the relationships between the terms of such a statement, and their validity after transformation into other forms; he is a specialist in grammar and syntax, not a literary critic.

Through this power of logical transmutation, the results of pure mathematical research are of immense importance in the 'natural' sciences. From the mathematical archives may be drawn theorems justifying unsuspected equivalences, computational short-cuts, generalized schemes of manipulation and powerful formalisms, that profoundly affect the contents of scientific knowledge. But there is no foundation for the self-serving prophecy that all pure mathematics will eventually find realization in empirical scientific knowledge – nor need the mathematicians apologize, concerning any particular result, that it does not seem immediately 'applicable'.[14]

The *applied* mathematician, on the other hand, is a sort of interpreter. He takes the messages flowing into the scientific archives, and combines and manipulates them into new forms, to be transmitted eventually to other scientists. He does his work best if, like the scramblers and descramblers on a military telephone link, he does not make nonsense of the original message. Unfortunately, because of his esoteric technical skill, he may flatter himself that as a professional *theoretician* he is the primary fount of human knowledge. Like the staff officer at headquarters he may conceive splendid schemes that ignore the realities of service in the front line of experiment and observation. In the physical sciences, where mathematical reasoning must be given its head, this sort of professional specialization and division of labour between bench experimentation and office calculation is inevitable; nevertheless it can, by academic snobbery and vanity, become a serious source of misunderstanding and error.

Mathematical reasoning is immensely more powerful than plain language when it comes to generating verifiable predictions, unpalatable conclusions, or unsuspected connections between known facts. This power is evident in two ways. By the *synthetic* or *euclidean* method,

[14] René Thom once said in a lecture 'Just as, when learning to speak, a baby babbles in all the phonemes of all the languages of the world, but after listening to its mother's replies soon learns to babble in only the phonemes of its mother's language, so we mathematicians babble in all possible branches of mathematics, and ought to listen to mother nature in order to find out which branches of mathematics are natural.'

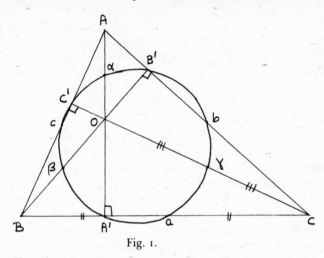

Fig. 1.

building upon precisely formulated axioms, and taking immense care
in the proof of each theorem, the mathematician can construct a
remarkable edifice of logical consequences. Anyone who has studied
elementary plane geometry, exploring, step by step, the properties of
straight lines, triangles and circles, must have been delighted to come
upon the 'theorem of the nine points circle'.[15] Yet who would believe
such an astonishing proposition (Fig. 1) on the basis of empirical
verification of a few particular instances or rather vague 'hand-waving'
of the kind that is familiar in popular scientific lectures?[16]

The analytical or cartesian method exploits algebraic symbolism as
an instrument for automating speech; formal logical operations can
thus be followed through in much greater depths than can be managed
by the unaided human brain (5.8). Thus, for example, the meaning
of a perfectly grammatical sentence, such as, 'The book the man the
gardener I employed yesterday saw left is on the table'[17] is almost
impossible to grasp verbally. Yet it is child's play to evaluate an
algebraic expression of similar structure, such as

$$[B+\{M+(G+I\cdot E)S\}L]I'+T$$

[15] 'In any triangle ABC, the midpoints of the sides (a, b, c), the feet of the perpendiculars
from the vertices (A', B', C') and the bisectors of the lines from the vertices to the
orthocentre (α, β, γ) all lie on the same circle.'
[16] 'I would like to make a distinction between a proof and a demonstration. A demonstration is a way to convince a reasonable man, and a proof is a way to convince
a stubborn man': M. Kac (1973) in *The Physicist's Conception of Nature* edited by J. Mehra
(Dordrecht: Reidel) p. 560. All scientists are, of course 'stubborn men'!
[17] Quoted from *Chomsky* by J. Lyons (1970: London: Collins).

by merely following the rules of algebra. The fascinating power of the digital computer rests upon its ability to perform an immense number of elementary manipulations of this kind, in pursuit of the logical consequences of some symbolic formula (6.10).

The essence of mathematical reasoning is that it is perfectly transparent in thin sections, yet intellectually opaque in bulk. A sequence of geometrical theorems or algebraic manipulations cannot be comprehended as a single mental argument; for a long calculation, a computer becomes a 'black box', whose inner workings must be trusted even though they cannot, in practice, be followed through from beginning to end. This lends to the results of a complicated mathematical calculation a peculiar novelty that can give immense prestige and rhetorical force to a successful prediction (2.8).

Suppose, for example, that we have observed one man, and another man, and another man going into an empty room; there will be little surprise at the subsequent confirmation of a prediction that the room contains three men. On the other hand, suppose that we refer to the 'mathematical theory of continuous groups' and point out that the 'octet of irreducible representations of the SU(3) symmetry group is incomplete', and hence that there should be an eighth elementary particle of a particular type;[18] it savours of magic, and is extraordinarily compelling towards acceptance of the theory, when just such a particle is discovered. Logically speaking, these two cases of scientific prediction are almost equivalent; rhetorically, they are poles apart.[19]

It is interesting to note, however, that experience with the application of the theory of continuous groups has taught theoretical physicists to 'see through' such calculations, so that nowadays the logic of such a transformation of the observational data has become almost as obvious as counting sheep (5.4). As science evolves, new theoretical models, newly discovered phenomena, new experimental techniques and new mathematical methods become so familiar that they are incorporated into the 'world map' that every scientist carries in his head (4.4). Deductions whose confirmation would once have seemed quite astonishing become entirely routine, as if no more than elementary exercises in mental arithmetic or 'physical thinking'.

The *creative* role of mathematics in science is balanced by its use as

---

[18] This refers to the famous prediction in 1961 by M. Gell-Mann and Y. Neeman of the existence of the 'Omega-minus' particle, which was discovered in 1964.

[19] But if the first prediction had been 'falsified' by there being found only two men in the room, we might say that we had evidence of a cannibalistic murder; whereas failure in the second case would just be a sad blow to another elegant but over-optimistic theory!

an instrument of *criticism*. For successful intersubjective discourse, various principles must already be agreed between the participants. Communication is fruitless between scientists that do not largely share a categorial framework (2.3). The very idea of striving towards consensus implies that it must already be achieved in some respect – that certain issues are no longer considered worthy of question. For example, 75 years ago, physicists who were interested in the laws that might apply in the atomic or cosmological scale could all agree that ordinary 'macroscopic' phenomena were described almost perfectly by the classical mechanics of Newton. The development of relativity theory and of quantum mechanics would have been impossible without agreement on this basic issue.

But the logical or material consistency of tacitly agreed principles, and the place of new principles amongst them, is often very difficult to determine without appeal to elaborate mathematical analysis. By 1900, classical mechanics had been subject to such an analysis for two centuries. This is a typical example of an *axiomatization programme*, to which many distinguished applied mathematicians (e.g. Lagrange and Hamilton) had devoted their talents. Such a programme can develop into a branch of pure mathematics; the goals of research move away from physical understanding towards problems such as finding a minimal set of axioms for a theory, or showing that the accepted principles are mathematically equivalent to some other apparently unconnected structure.

To the imaginative theoretical physicist, axiomatization seems a formal and sterile activity. But there is much to be learnt from the mathematical metaphors that are sometimes uncovered in this process – that the behaviour of an electron in an atom is 'like' the vibration of air in a spherical container, or that the random configuration of the long chain of atoms in a polymer molecule is 'like' the motion of a drunkard across a village green. It is wise to recall that the mathematical foundation of Einstein's theory of general relativity was Minkowski's demonstration that the equations of special relativity could be formally represented as an imaginary rotation of axes in an abstract four-dimensional space-time continuum – a representation that seemed to have nothing to do with the physics of measuring rods, clocks and beams of light.

2.5 Metaphors and models

Axiomatization is the final, decadent stage of *theorizing*. A body of quantitative scientific knowledge that has been assimilated to an abstract structure of mathematical relations is no longer fit for human consumption; it provides fodder only for the computer (6.7). Nothing more can be done with it; it has scientific interest only as an instrument for the advancement of learning in other fields. This stage of maturity is exemplified by classical mechanics and thermodynamics, which are no longer questionable in themselves, but which are the foundations for such extraordinary feats of computational virtuosity as the guidance of space vehicles or the prediction of the weather.

Mathematical theorizing is equally unprofitable in the primitive, exploratory phase of a new branch of science (7.3). An elaborate, sharply defined theory conjured up out of fragmentary and uncertain evidence may have its intellectual charms, but can prove a misleading guide to research. This is probably the situation in the field of brain research, where the cerebration of mathematical theories of memory, perception, cognition, etc. has had little impact on the experimental study of the subtle and complex facts. At this stage, the quantitative representation and mathematical transformation of the data is essentially *phenomenological*; correlations and regularities are noted, for themselves, without reference to more general systems of explanation.

Eventually, however, the innumerable messages that pass to and fro between the observers and into the archive must be categorized according to some general ordering principle. Scientific knowledge is incomprehensible – i.e. cannot be grasped by the human mind – unless it can indeed be represented by a few relatively simple and coherent theories. In constructing and manipulating mathematical theories we rely very heavily upon models – so much so, that this word has become a (slightly trendy) synonym for a 'theory' in all branches of the natural and behavioural sciences.[20]

The mathematical models of theoretical physics have been widely studied by historians and philosophers of science. As products of the human imagination, and dominant symbols of successive revolutions of thought, they are of immense intellectual significance. What is the source of their rhetorical power and authority?

[20] The defect of this fashionable usage is that it often lends to a vague hypothesis an unwarranted air of concreteness and logical consistency. The verb 'to theorize' is now conjugated as follows: 'I built a model; you formulated a hypothesis; he made a conjecture.'

By definition, a model is not a complete and faithful rendering of reality. It is no more than an analogy or *metaphor*. It implies a structure of logical and mathematical relations that has many similarities with what it purports to explain, but cannot be fully identified with it. The wise theorist does not assert or attempt to prove the necessary validity or verisimillitude of his model; this is to be discovered by further experience. He says (often in just these words) 'Suppose we think of it this way: what follows?' Even Kelvin's mechanical model of the ether as a medium packed with levers and pulleys, for all its apparent concreteness, cannot have been meant as a serious *description* of empty space. At its conception, a model is no more than a guide to thought, or a framework for a mathematical interpretation of inexplicable phenomena.

What is not always recognized is that a model, being drawn from another field of knowledge than that to which it is to be applied, carries a certain amount of pre-existing understanding of its own properties. The Rutherford–Bohr picture of the atom as a planetary system of electrons in orbit around a nucleus owes its strength not only to the basic principles of classical physics, but also to our familiarity with just such systems in astronomy. The knowledge conveyed in such a metaphor goes much further than the technique of solving a few differential equations; it contains a large element of experiential and intuitive understanding (5.4). To exclude such elements by axiomatization of the model in advance of its application would be sterile, for it would put off the occasion to discover the insights about the *explicandum* for which the hypothesis was originally proposed. It would be difficult, even now, to give a precise logical definition of Darwin's model of interspecific competition as the motive force of organic evolution. This model derived its explanatory power from the fact that its audience were familiar with industrial and social competition, of which many characteristic features could be grasped and compared with biological phenomena without formal demonstration or proof.[21]

These tacit features of a well-conceived model both restrict and enlarge its capabilities. The restriction is advantageous, for it spares us from self-contradictory conjectures. The fact that the model system

[21] This property of a model exemplifies a more general principle suggested by Michael Polanyi (1958) in *Personal Knowledge* (London: Routledge & Kegan Paul) p. 169: 'These powers enable us to evoke our conception of a complex ineffable subject matter with which we are familiar by even the roughest sketch of any of its specifiable features. A scientist can accept, therefore, the most inadequate and misleading formulation of his own scientific principles without ever realizing what is being said, because he automatically supplements it by his tacit knowledge of what science really is, and thus makes the formulation ring true.'

is actually realized in its original sphere implies that its defining properties are coherent and mutually consistent. When, for example, Niels Bohr set up a mathematical theory of nuclear fission, using as his model the break up of a drop of liquid, he did not need to explore his equations in detail to prove that their solutions were unique, mathematically stable, etc.; these properties could be taken for granted from his familiarity with the physics of real liquids. A theory constructed thus, by mathematical analogy, has a 'realizable coherence' which can be grasped intuitively long before it can be proved. The defect of many a theory built around a hypothetical mathematical relationship – for example, the non-linear field equations to which Heisenberg gave the later years of his life – is that one simply does not know in advance whether it will 'work' at all. All too often, the 'interesting' properties for which it was conceived are accompanied by pathological mathematical and physical features which make it meaningless in practice. By confining his imagination to realizable models, the theorist avoids speculative schemes that fail through sheer inconsistency and automatically keeps within the bounds of consensibility.

On the other hand, the imagination is enlarged by a model that is known by experience to exhibit a rich variety of phenomena. When, for example, the distribution of stars in a galaxy is likened to the distribution of molecules in a gas, the analogy may be intended to apply to some very smooth and simple processes of uniform expansion or steady flow. But as the model acquires scientific status and authority, it suggests the possibility of astronomical phenomena akin to turbulence, or wave propagation, which are familiar properties of ordinary gases. Such 'physical' properties, known so well from experience, would suggest themselves long before the corresponding mathematical solutions would have been found in the equations of motion of the galactic system.[22]

In the search for consensus, a 'realistic' model also has major advantages over more abstract theoretical schemes. Precisely because of its internal consistency and tacit limitations, it is more dramatically falsifiable in some vital particular. The eighteenth-century model of heat as an 'imponderable fluid' was effectively falsified by the dynamical production of heat in apparently endless quantities out of the

---

[22] Once again, Polanyi (*Personal Knowledge*, p. 104) states the general principle '[These] major intellectual feats demonstrate on a large scale the powers which I have claimed for all our conceptions, namely of making sense beyond any specifiable expectations in respect to unprecedented situations'.

same piece of machinery. An aspect of the caloric theory that might have been left vague in some abstract mathematical formulation – that it was a 'substance' that could not be created 'out of nothing' – was brought into the open by the very concreteness of the image conjured up by the model. Again, the Michelson–Morley experiment showing that the velocity of light seemed independent of the velocity of the observer was much more serious for the model of light as a 'vibration of the luminiferous ether' than it was for the electromagnetic theory in which light is just a wave-like solution of Maxwell's equations, which Lorentz could easily modify to give the correct experimental result.

From these characteristics, we may easily judge the power that a successful theoretical model may acquire over the human mind, and need not be surprised that most of the mathematical theory of physics has this sort of origin. There are very few episodes comparable with Dirac's relativistic theory of the electron, which was hypothesized directly as a mathematical equation without previous physical analogies. In some cases, of course, the metaphorical origins of mathematical principles may have become hidden with historical change (6.7). In the modern theory of elementary particles, various 'symmetry' properties are assigned to the different particles to explain their behaviour. For all practical purposes, these properties are defined through certain algebraic relations, which constitute, in effect, the whole content of the theory. Yet this formal mathematical theory did not arise by mere cerebration; it is no more than an inspired generalization and abstraction of the mathematical description of a perfectly straightforward physical model – the effect of changing the direction of the axis of rotation of a spinning object.[23]

Generously extending the concept of a model in this way, we can scarcely avoid the conclusion that most of our thinking is analogical and metaphorical.[24] It is certainly true that the precise, unambiguous, 'mathematical' communications that are ideal for intersubjective

---

[23] 'It almost seems as though, for the formalist programme to work at all, a previous stage of science making use of models [e.g. classical physics] is necessary in order that a sufficiently complex observation language shall have been built up.' Mary Hesse (1963) in *Models and Analogies in Science* (London: Sheed & Ward).

[24] 'We sometimes use metaphors as devices for explaining theory when they were actually essential to the formation of theory. In discussing a theory of genes the lecturer may say "Think of it, if you will, as a kind of code", when, in fact, he has no other way of thinking of it. Something like this happens when scientists teach theories of submicroscopic phenomena in terms of mechanical models, or when biologists teach theories of natural selection in terms of models of human competition' D. A. Schon (1963) *Displacement of Concepts* (London: Tavistock) – see also W. H. Leatherdale (1974) *The role of Analogy, Model and Metaphor in Science* (Amsterdam: North Holland).

comparisons and consensus building get organized around relatively compact theoretical models that come to constitute the cognitive contents of such a science (4.4). This is about as far as we can go by attending to the language and logic of the messages that scientists send one another or store in their archives; we must now try to say something about the 'world' that these messages purport to describe.

2.6   *The logic of experience*

The categories and logical relations that govern the form and content of consensible communications are supposed to reflect corresponding categories and relations in the reality they describe. This we take for granted. Yet, as G. H. Mead so eloquently puts it,

> in the world of immediate experience, the world of things is there. Trees grow, day follows night, and death supervenes on life. One may not say that relations here are external or even internal. They are not relations at all. They are lost in the indiscerptibility of things and events, which are what they are. The world which is the test of all observations and all hypothetical reconstructions has in itself no system that can be isolated as a structure of laws or uniformities, though all laws and formulations of uniformities must be brought to its court for its *imprimatur.*[25]

This is one of our ultimate problems, to which we must continually return. Grant for the moment relatively efficient procedures of observation and experiment, yielding information whose basic consensibility is not in doubt (5.6). Grant that we can make quite reliable quantitative measurements, on whose results we may operate by sound mathematical techniques. We still cannot assert that scientific knowledge, embodied in well-articulated theoretical models, is a faithful representation of the real world.

The fundamental difficulty is that the *logic of empirical statements is not the logic of mathematical theory.*[26] Statements about the real world are always subject to uncertainty. They cannot all be given precise status – 'true' or 'false' – their logic is *three-valued*, falling into the categories 'true', 'false' and 'undecided'.

Suppose, for example, we·say, 'the distance from London to Bristol is 120 miles'. What do we mean by 'London' and 'Bristol'? Does 'London' mean the City of London, or the area covered by the Greater London Council? Is Avonmouth a part of Bristol for this

[25] *The Philosophy of the Act* (1938: University of Chicago Press), p. 31.
[26] Stephan Körner (1966) *Experience and Theory* (London: Routledge & Kegan Paul).

purpose? Does the measurement refer to the distance between the centres of the areas of the two regions, or to certain conventional points such as the Post Office Tower? Even if these points were defined, could we determine the distance to an accuracy of 10^{-8} cm? Does the measurement allow for thermal expansion, the microscopic movements of the earth's surface, the influence of the tides, the phases of the moon, the behaviour of the local population and other conceivable interfering factors? The point is that this sort of statement – or any other empirical statement – could not be made indefinitely precise. We would eventually be forced to admit that a statement of the form, 'the distance from London to Bristol is less than (say) 193.6142857 km,' is undecided, and cannot be taken to be true or false. This objection applies to any measured quantity such as the mass of an electron or the magnetic field of the sun. As Körner proves in detail, this is a limitation in principle, not a mere practical difficulty. And the same objection can be made against any sharply defined categorial scheme for the description of reality, whether 'quantitative' or 'qualitative'.

On the other hand the prime requirement for consensibility of a scientific communication is that it should be unambiguous (2.2). The mathematical transformations to which we propose to subject it will assume that it satisfies ordinary *two*-valued logic (2.3). In setting up a mathematical theory, we inevitably idealize the concept of a measured distance, for example, as the value of a continuous variable, represented technically as a limit point or 'Dedekind cut' along the sequence of rational numbers – etc. etc. By definition, the undecidable case is excluded – not for subtle reasons to do with the foundations of mathematics and Gödel's theorem, but as a practical necessity in constructing and testing workable theories.[27]

From this it follows that the *identification of ideal with empirical statements is not deductive*. Having neglected the uncertainty in our premises, we can never be sure of the logical necessity of our conclusions. Every theoretical calculation becomes metaphorical; it may *depict* reality (4.3) but cannot *mirror* it.[28]

[27] We could, of course, construct a complete mathematical system founded on three-valued logic, and thus keep empirical and theoretical statements on the same footing. But this would be at the expense of *human* intelligibility: *our* science is not like that, although Black Clouds might have the capacity to comprehend reality that way (5.10). And in what sense would it be more credible or reliable than our present science?

[28] 'If, as Condillac says, it is true that "all science is a well-made language", it is no less true that all natural phenomena constitute a badly understood language. Also consider Heraclitus "The Lord whose oracle is at Delphi neither speaks nor conceals, but gives signs".' René Thom (1975) *Structural Stability and Morphogenesis* (Reading, Massachusetts: Benjamin) p. 117.

In the physical sciences, this objection, although a large nail in the coffin of doctrinaire positivism, is not of great practical significance. But when we go beyond biology to the behavioural and social sciences (7.2) it is devastating. The two-valued logic that is ascribed, unconsciously, to idealized categories ('function', 'role', 'intelligence') does such a grave injustice to their inherent properties and behaviour as to make nonsense of symbolic, logical, mathematical communication about them. The most elegantly articulated and computationally complex model of such phenomena is no more reliable or persuasive in representing or getting at 'the truth' than the logic chopping of a mediaeval scholar.

2.7 *Physics and physicalism*

How, then, is it possible that in *physics* we are able to use mathematical reasoning with the utmost confidence? What feature of the *physical sciences* – astronomy, chemistry, geophysics, metallurgy, materials science, engineering, etc. – guarantees this power?

The answer is, simply, that these sciences are created by systematically following the strategy implicit in the definition of science as 'The Art of the Soluble'.[29] They are deliberately developed to exploit mathematical methods. Physics, we say, makes use of quantitative observations; in fact, *only* quantities that can be represented numerically and transformed mathematically are permitted in the physical sciences. It is not simply good fortune that physics proves amenable to mathematical interpretation; it follows from careful choice of subject matter, phenomena and circumstances. Physics defines itself as the *science devoted to discovering, developing and refining those aspects of reality that are amenable to mathematical analysis.*[30]

It is easy to see how this strategy works in practice. To bridge the logical gap between the empirical and the ideal (2.6) we search for categories of experience where the *tertium quid* – the undecidable middle ground – is as small as possible. In this way, we hope to avoid any gross errors when we represent reality by theories grounded on two-valued (i.e. mathematical) logic.[31] Consider the typical subject matter and concepts of physics:

[29] P. B. Medawar (1967) *The Art of the Soluble* (London: Methuen).

[30] 'Physical science may be defined as "the systematization of knowledge obtained by measurement". It is a convention that this knowledge should be formulated as a description of a world called "the physical universe".' (A. S. Eddington).

[31] 'A possible explanation of the physicist's use of mathematics to formulate his laws of nature is that he is a somewhat irresponsible person. As a result, when he finds a connection between two quantities which resembles a connection well known from

Atoms and electrons provide us with identical, distinct *countable* objects.

Space and time approximate to *continuous variables.*

Mass and charge are found to be *invariant* or *conserved parameters.*

Stars and crystals are *simple geometrical forms.*

Velocity and force are *linear vector quantities.*

Electricity and magnetism are *vector fields.*

Planets and stars are *weakly interacting systems* – and so on.

All these qualities, which we think of as fortunately provided by nature for our better comprehension of its glories, have been selected by us, in order that we may make some progress in representing them in terms of ideal models. If stars were really the spiky objects depicted in Christmas decorations, then astrophysics would scarcely exist and astronomy would be classed as a descriptive science along with geology and botany. If nuclei up to mass one million were stable, then the properties studied in atomic and nuclear physics might be supposed to be as arbitrary and incalculable as those of the innumerable molecular species of organic chemistry.

The supposed goal of physics – the search for the *fundamental laws of nature* – exemplifies this strategy. Given the messy, chaotic world of everyday things, the physicist applies his peculiar methods to distill out the mathematically consensible essences. He extracts algebraically simple qualities, such as mass and spatial extent. He deliberately breaks things into 'elementary' parts, of greater simplicity: the *organism* is divided into *cells*; each cell is analysed into its chemical *molecules*; each molecule is broken into its constituen *atoms*; the atom is separated into *electrons* and *nucleons* – and so on. At each stage, the invariance and indistinguishability of the elementary parts increases, so that the possibilities of a mathematical description of their properties and phenomena become wider and more inclusive. In other words, he follows the prescribed career of the specialist; he learns 'more and more about less and less', so that by the time he reaches the elementary particles he knows 'everything about nothing'.

In his experiments, the physicist contrives to create conditions that are simple to observe and analyse; he makes single oil drops, each carrying a small number of electrons, or causes collisions between two energetic particles of known type. He constructs artificial systems

mathematics, he will jump at the conclusion that the connection is that discussed in mathematics simply because he does not know of any other similar connection'
E. P. Wigner (1969) in *The Spirit and Uses of the Mathematical Sciences* edited by T. L. Saaty and F. J. Weyl (New York: McGraw Hill) p. 131.

with nearly ideal properties; pure, perfect crystals, rarefied gases, homogeneous magnetic fields, beams of mono-energetic particles. And by going to extremes – low temperatures, high pressures, strong fields, high energies – he extends the conditions of existence of his chosen material so as to explore the phenomena of their interaction over the widest range of the variables, thus checking his theoretical models and their mathematical relations outside their normal circumstances.

The extraordinary accuracy and remarkable insights obtainable, through 'the unreasonable effectiveness of Mathematics in the Natural Sciences'[32] are not necessarily due to 'God being a mathematician'. They are associated with scientific communications in logico-mathematical language, directed towards the construction of consensual models governed by two-valued logic. Physics is the harvest of this Kantian fisherman, whose net only catches fish larger than the size of its mesh, and who proudly proclaims as a 'law of nature' that all fish are larger than that size.

There is nothing fundamentally wrong with physics as such; but it is an inappropriate model for *all* consensible and potentially consensual knowledge. As we shall see there are other reliably consensible forms of intersubjective communication than those trimmed to the logic of mathematics (3.2). In treating biological and social phenomena, where the middle ground of three-valued logical categories cannot be ignored, one must not fall into the doctrine of *physicalism*, in which it is asserted that all scientific knowledge should be, or should strive to be, or will eventually be, just like modern physics. This is merely a specialized and naive form of *scientism*, which holds that all reliable knowledge of the world and its workings should be, or should strive to be, or will eventually be, 'scientific' in some ill-defined sense.

2.8 Prediction

In spite of its privileged use of mathematical language, physics is not all that easy. Theoretical models must be imagined and tested before they become part of the consensus. The psychology of invention lies outside our chosen field of study; but we are obviously keenly interested in the critical assessment and validation of theoretical conjectures, hypotheses and models. The reliability of knowledge in the physical sciences depends less, in practice, on the validity of basic principles and standard methods than it does on the extent to which

[32] E. P. Wigner (1969) *The Spirit and Uses of the Mathematical Sciences*, p. 123.

a new theory has been thoroughly tested for agreement with experience (6.5). The fact that the uncertainty of an empirical statement can be minimized by careful quantitative measurements under precisely controlled conditions does not inhibit speculative theory-building: the active research literature of physics is well supplied with fanciful conjectures for which there is little evidence and which are later tacitly dropped overboard.

The consensus principle implies that the validation of theoretical models must be fully public, and cannot be accepted on the authority of any one scientist. The essence of a mathematical model is that it is perfectly consensible, and unambiguously defined, so that tests of conformity with experiment are relatively precise, and can be carried out by anyone (3.1). This avoids the danger (encountered in some branches of science and scholarship) that the skill and insight required for such tests is only available amongst those persons who are already to some extent committed to the theory in question.[33]

Needless to say, the most impressive way of validating a scientific theory is to confirm its *predictions*. This is especially convincing in the physical sciences because a prediction expressed in mathematical terms is relatively unambiguous and yet can seem uncannily inconsequent on the premises from which it is derived (2.4). But the persuasive power of a successful prediction arises from the fact that it could not have been deliberately contrived. The most famous examples, such as Mendel'eef's prediction of the existence and properties of undiscovered elements to fill the gaps in the Periodic Table, or Gell-Mann and Neeman's prediction of the Omega-minus particle to complete an $SU(3)$ octet, have astonishing rhetorical power. Without trying to assess the tiny *a priori* probability that a random prediction is likely to be confirmed, we come rapidly to a feeling of certainty that the theory on which the prediction is based must be valid.

But suppose that Mendel'eef had secretly discovered 'eka-aluminium' *before* he proposed his scheme of the elements – or imagine that both Gell-Mann and Neeman had been scrutinising bubble-chamber photographs and had observed some Omega-minus tracks *before* they set up the $SU(3)$ classification. If this were known to us, then we should be much less astonished at their 'predictions', and should have had much less faith in their theories. Logically, this

[33] Note Rutherford's method of observing scintillations in his experiments on radioactivity: the counting of scintillations was done by undergraduates who did not know the purpose of the experiment: the curves were drawn by persons who did not know what results were expected. P. L. Kapitza (1973) in *The Physicist's Conception of Nature* edited by J. Mehra, p. 757.

makes no sense; the question whether an experimental fact that fits a theory was previously known is quite irrelevant. *Psychologically*, however, it is of the utmost significance, demonstrating that inter-subjective communication and persuasion are key factors in the machinery of science.

It can be argued, indeed, that the fundamental purpose of science is to acquire the means for reliable prediction. From a purely practical point of view, if we are not simply gathering knowledge for its own sake, then we are doing so for guidance in decision and action. Since the past is beyond human influence, we must be seeking some comprehension of the *future*, as it will be or as we should like to make it. Science is to be believed (5.5), science is to be regarded as reliable, in so far as it actually supplies us with just that power.

We turn to science, also, as an extension of the psyche in its conscious, rational dimension. Through the mental faculties of consciousness and rationality, by means of memory and imaginative forethought, we humans live beyond the present moment, and carry within us personal segments of time, from the past, into the future. By interpersonal communication, by interaction with socially-stored knowledge, we may each enormously extend these segments, from the distant cosmological or historical past to the potential triumphs or disasters in the shape of things to come. In science, therefore, we seek knowledge that transcends the barriers of here and now, whether through postulations of perfectly invariant Laws of Nature, or through rational reconstructions or pre-constructions of past or future events.

Not all topics about which we communicate socially have this concern with the future. Law, for example, deals in justice, which is timeless, applied to events of the past. Politics (where 'a week is a long time'!) seldom escapes from the demands of expediency. Poetry is a commentary on the human condition, to be discovered and rediscovered by each of us, in his own generation. But scientific knowledge could almost be defined as what remains of our picture of the world after continual projection and reprojection into the future, and is moulded into that form by the selection we make of messages from our fellow scientists.

The obsession of the philosophers of science with the problem of *induction* is thus quite intelligible. Without some metaphysical principle of 'permanence', of 'order', of 'lawfulness', of 'continuity' linking one epoch of time with the next, the whole enterprise is quite pointless. It is obvious, however, that such a principle must lie outside the whole categorial scheme that it is called upon to dominate. That is why a demonstration of a successful scientific prediction is not especially compelling in technical logic, but speaks nevertheless right to the heart of the scientist or other living person. We want our science because we want to know how things will turn out; we are impressed by science and believe in it when it does just that. The justification is circular, but it is hard to see a way out of it in this world!

It must be emphasized, however, that the well-known historical cases of the confirmation of theoretical predictions do not imply that the physical sciences have perfect predictive power. Prediction is a very general term, with diverse levels of presumed capability. At one end we have Laplace's fantasy of a superhuman computer who might predict the whole future from the present state of the universe, atom by atom. More realistically, we might hope to foretell the future behaviour, for a certain period, of a particular complex system, or define a range of probable future states for an ensemble of such complexes. In practice, we are usually delighted if we are successful in predicting the existence, as yet unverified, of characteristic categories of nature, such as quarks and black holes, and are well satisfied if we can calculate the trajectory in space and time of a hypothetically simplified model system such as a planet or a spacecraft. We believe in science because we occasionally verify remarkable predictions based on scientific theory; we have no warrant for a belief in theoretical predictions because of the occasional success of science.

2.9 *The fit between theory and experiment*

The practising scientist rarely has an opportunity to make a definite theoretical prediction that is subsequently confirmed by experiment. Theories are usually validated by much less compelling evidence. In physics the normal test of a theory is simply that it should yield results that 'fit' the experimental data. That is to say, it should be possible to deduce from the model, using well-founded principles, good quantitative formulae for measurable physical properties. The more precise the numerical agreement between theory and experiment, the more convincing becomes the model upon which the theory is based.[34] Thus, the Rutherford–Bohr 'planetary' model of the hydrogen atom (2.5) was immensely compelling, not because it made any predictions of previously unobserved quantities, but because it gave uncannily ac-

[34] 'The future truths of Physical Science are to be looked for in the sixth place of decimals.' A. A. Michelson (1902) *Light waves and their uses* (University of Chicago Press).

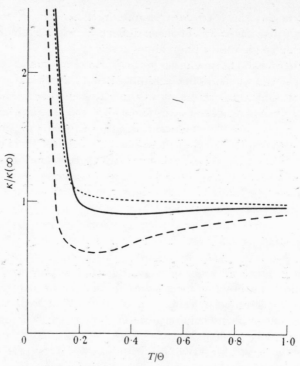

Fig. 2. My new theory of thermal conductivity of sodium (. . .) was a better fit to the experimental data (——) than the previous theory (----), wasn't it? From Ziman, J. M. (1954). *Proceedings of the Royal Society*, Series A.

curate values for the optical spectrum of this element, which had been known for a quarter of a century. Similarly, in the 1950s, a dubious mathematical technique of getting finite answers by subtracting infinite quantities from one another was made respectable as a theory of quantum electrodynamics by giving a very precise value for a known anomaly (the Lamb shift) that had been predicted in principle and carefully measured a few years before.

But even these historical examples are misleading. It is seldom that the numerical fit between theory and observation lies within the instrumental uncertainty of the experimental technique. In practice, the theoretical physicist may have no better evidence for his model than that it gives the correct qualitative pattern of results for a variety of data, none of which are closely fitted. This is often represented graphically. A diagram is published (Fig. 2) showing that the theoretical curve has the right 'trend', or that it exhibits recognizable 'features', such as 'peaks', or sharp discontinuities that can plausibly be identified with known physical phenomena.

This fuzziness is disappointing for the convergent-minded lad who goes into physics looking for certainty of proof and belief; but it is not really surprising. Calculable mathematical models are not only embedded in an over-precise two-valued logic; they also lack too many of the complicating and confusing features of real life to 'predict'[35] the outcome of an actual observation. In other words, even in physics, despite all our efforts to choose situations where the undecidable 'third category' of empirical statements is of minor significance (2.7), our theoretical models are usually idealizations that cannot be in very close agreement with the whole truth.

For this reason, Karl Popper's famous criterion[36] for an acceptable scientific theory – that it should, in principle, be *falsifiable* – is strategically sound but tactically indefensible. It turns out, in practice, that almost every theory is to some extent 'falsified' by the relevant observations: the question then hinges on whether this failure is to be treated as a genuine objection, or whether, pending conceivable improvements in formulation or computation, it may be temporarily overlooked.

This point is well illustrated by a scientific controversy, that has not yet been resolved – the anomaly of the 'missing solar neutrinos'.[37] Briefly, the accepted theories of nuclear reactions in the core of the sun predict the generation of a large flux of neutrinos, which should be observable as they pass through the earth. The neutrino is a very, very elusive particle, but it can be observed, very, very occasionally, by very large, refined and expensive apparatus. Experiments of this kind, on a heroic scale, have not confirmed the theory; the flux of solar neutrinos is not precisely measurable, but in all experiments appears to be very much less than the theoretical value. But what has been falsified? Even if we accept the experimental results at their face value, this does not necessarily mean that the theory of nuclear reactions is all wrong. The calculations make many assumptions, such as the rate

---

[35] Significantly, the word is commonly used amongst physical scientists to refer to any quantitative deduction from a theoretical model that can be compared with experiment, *whether or not this experiment has yet been carried out.*

[36] *The Logic of Scientific Discovery* by Karl R. Popper (1959: London: Hutchinson). Popper proposed this demarcation criterion for science in the 1930s, but the same point had been made in 1924 by J. Nicod (English translation 1969) *Geometry and Induction* (London: Routledge & Keegan Paul) p. 190.

[37] See, for example 'Neutrinos from the Sun' R. B. Kuchowicz (1976) *Reports on Progress in Physics*, **39**, 291–344.

of mixing in the solar interior, or the uniformity of conditions over long periods of time, that are difficult to decide independently and that strongly affect the results. Some astrophysicists interpret the anomaly as a radical falsification of the theory, and are looking for new basic models of the solar energy source; others, with equal justification, estimate the consequences of plausible modifications in the details of the conventional model, in order to 'explain away' the anomaly.[38] Whatever the outcome of this episode, which vividly illustrates the turmoil of imagination, scepticism and criticism, and the dynamical interaction between theory and observation, in this branch of science, only the hindsight of armchair theorists will eventually explain to us what was really being falsified and why.

In other words, the validation of a theoretical model by appeal to experiment is not a mechanical process whose outcome can be determined by simple logic (6.5). It hinges on the expert judgements of physicists, who must decide for themselves whether there is an *adequate* fit between theory and experiment, given the uncertainties of the data and the unavoidable idealizations of the mathematical analysis. The skill to make such judgements comes from experience.[39] The theoretical physicist may rely, for example, on his unformulated estimate of the very small likelihood that any other simple rational model could have given quantitative results as close to reality as the one being tested. On the other hand, he may have a shrewd suspicion that an apparently significant formula derived from an elaborate model may be no more than a simple consequence of dimensional arguments which would hold for almost any theory in the same circumstances. Or, faced with claims for a qualitative pattern of agreement with experiment, he may exercise his familiarity with the behaviour of various types of mathematical system to assess intuitively (5.4) the possible range of 'features' that might appear on the theoretical curve, and the extent to which they might be brought into good quantitative agreement with observation by permissible adjustments of the parameters of the model.

The standards by which a physical theory is judged are not fixed, but depend on the nature of the subject, and its state of development.

[38] 'The critical problem is to determine whether the discrepancy is due to faulty astronomy, faulty physics, or faulty chemistry.' V. Trimble and F. Reines (1973) *Reviews of Modern Physics*, **45**, 1–5.

[39] This is the main theme of Polanyi's philosophy of science, in *Personal Knowledge*. The falsification criterion has been thoroughly discussed, logically, empirically, critically and wittily by A. Naess (1972) *The Pluralist and Possiblist Aspect of the Scientific Enterprise* (London: Allen & Unwin).

A highly speculative model whose results are within an order of magnitude of fragmentary observational evidence may be regarded as quite plausible in astrophysics or cosmology, but would not be taken very seriously in, say, the theory of semiconductors, where carefully controlled experiments could easily check the details. In the initial stages of a new branch of science, where little is known for sure, an imaginative conjecture may be valued as a guide to further research, however slight its basis. Once more, it is part of the professional skill of a research worker to balance, or oscillate, between faith and scepticism (6.3) as he threads his way to a true understanding. If he remains critical and disbelieving, he may reject an important new insight; if he swallows credulously every fantastic proposal, he will build no more than a theoretical house of cards.

Like Alice's Red Queen, who could 'believe six impossible things before breakfast', it may even be necessary to accept logically contradictions of models of the same system. In the theory of atomic nuclei, the essence of the 'liquid drop model' (2.5) is that the protons and neutrons within the nucleus are packed so densely together and repel each other so strongly that they form something like an incompressible fluid. This model is beautifully validated by phenomena such as nuclear fission, where the drop becomes unstable and divides in two. On the other hand, to explain the stability and radioactive properties of a large number of nuclei, the so-called 'shell model' has to assume that each proton or neutron can move with almost perfect freedom inside the whole volume of the nucleus, almost as if the other particles did not exist. This model, also, gives quantitative results in excellent agreement with experiment. It is easy enough to assert optimistically that these two models are merely idealizations or specialized aspects of a much more complex but unified theory which will eventually yield the correct description of the phenomena in all cases: but until, by some theoretical *tour de force*, this unification is actually achieved, the nuclear physicist is faced with manifestly contradictory models, each of which is so scientifically sound that it has earned a Nobel Prize for its authors and proponents.[40]

It may be wiser to accept, provisionally, both points of view, looking for evidence that might show how they can be reconciled, rather than to assume that one or other model has been falsified and

[40] This was certainly the situation in the 1950s. It may be that the theoretical unification has now been formulated. I could easily find out by asking the experts, or by reading a few review papers. But the subject is highly technical, and not very interesting in itself. The status of a theory changes with time, and must not be judged on its 'final' state.

37

must be deleted from the archives of scientific knowledge. This strategy may be essential in the behavioural sciences where a complete, unified model is an impossible ideal (7.6).

Nevertheless, the principle of ultimate falsifiability is a most valuable maxim in the practice of research. Although it does not fully demarcate science from non-science, it draws attention, once more, to the orientation of scientific knowledge towards the factual future. For to *falsify* means no more than to have made an *unsuccessful prediction* (2.9). In practical terms, a scientific theory that could not be falsified in principle would be quite useless as an instrument of prediction; all conceivable futures would be the same to it, and we should ignore it in our calculations. On the other hand, a new theory that makes very definite predictions concerning the observable future not only promises to be of great scientific value if it proves valid; it will also be easily falsifiable by the turn of events.

There is a temptation at this point to enter a somewhat formal game of assigning *a priori* probabilities to the outcome of tests of a scientific hypothesis, in the hope of 'measuring' the strength of success or failure. But within the unassessable universe of conceivable hypotheses or possible futures, all such probability concepts are unquantifiable, if not totally meaningless. It is enough, perhaps, to suggest that a falsifiable hypothesis assigns a high weight to only a few of many alternative futures and is thus subject to critical disconfirmation if it is already in other respects of low *a priori* probability. But that seems only another way of saying that a scientific theory is most convincingly validated by the successful testing of a sharp and apparently unexpected prediction – and can be relied upon very heavily if that should fortunately occur. In other words, the falsifiability criterion emphasizes the *quality* required of a successful confirmation of a prediction; but it could seem too negative, sceptical (5.6) and hypercritical towards what has been proved, beyond peradventure, worthy of strong belief.

2.10 *Validating physics*

Despite the encircling gloom along the way, the path of physics has carried mankind on to very firm ground in new territories of knowledge. If it is not the paradigm of all science (2.7), it certainly offers us the best examples of reliable knowledge achieved by the scientific system. The high consensibility of mathematical communication has undoubtedly facilitated an almost perfect consensus concerning precise theoretical models which adequately explain a vast mass of empirical data. In those areas of physics where this consensus has been achieved, theoretical systems such as those of classical mechanics and non-relativistic quantum mechanics have been thoroughly validated, beyond any shadow of doubt, by appeal to experiment and observation. What are the grounds for our unusual faith in the invincible product of so many fallible human intellects?

It is easy, in text-book fashion, to point to a few key episodes where a new theoretical model has been made particularly convincing by a confirmed prediction or by very close agreement with experiment. For rhetorical success in such cases, various criteria must be satisfied. Thus, the model must not be purely phenomenological, but must be based upon well-founded and independently validated principles. It should not contain a variety of adjustable parameters, or hidden variables, that have to be invoked to 'explain' discrepancies between theory and experiment. The theoretical properties of the model should be sharply defined, and derived with sufficient mathematical rigour to be compared unambiguously with the observed phenomena. To be interesting and convincing, the theoretically 'predicted' properties of the model should not be obvious logical or arithmetical consequences of its initial assumptions – and so on.

In principle, it is the duty of the theoretical physicist and his experimental colleague to collaborate in engineering such episodes, where an imaginative but calculable model is matched to an elegant but precise experimental measurement. In practice, such episodes are rare. Experimental data are accumulated with little guidance from theory, or speculative models proliferate without clinching evidence. The text-book experiment that supposedly validates the model may not, in fact, be carried out until long after the theory has become part of the consensus of the subject, or an elaborate theory, with many unnecessary complications, may become entrenched on the basis of inaccurate data.

The text-book description of a physical science fails to do justice to the *network* of interrelated models, experiments, concepts, mathematical techniques, instruments, materials, properties, etc. that constitute the corpus of knowledge in that science (4.2). Our confidence in any particular element of this science cannot be rested solely upon one or two other elements, but is deeply embedded in our consciousness of a multitude of related facts and opinions. Not all the elements of the network are of equal weight or credibility, but they must all be taken into account in an assessment of the reliability of our knowledge in that field.[41] Many particular consequences of a theory may seem far from the experimental facts, yet the picture as a whole may be completely convincing for its consistency and wide application. Physicists do not accept the Schrödinger equation of wave mechanics just because it happened (like the Bohr theory of planetary orbits) to give the observed spectrum of a hydrogen atom: wave mechanics now stands unquestioned because it could explain, more or less quantitatively, more or less successfully, almost all the properties of atoms, molecules and crystals.[42] When, for a time, it seems to fail to explain some unusual phenomenon such as superconductivity, we scarcely conceive that it has been falsified but assume that we have made a mistake in our calculations, or that we have misunderstood the physical situation.

But it is not our purpose in this book to explore all the byways of physics, or to expound the mysterious arts of the professional theoretical physicist. His personal experience of the validation of scientific theories is usually very sobering. He learns how easy it is to persuade oneself of the validity of a model which later turns out to be false, and comes to realise that even in very strongly mathematical and well-defined scientific issues it may take a long time, much criticism and the death of many promising conjectures (if not necessarily of their authors!) before a reliable theory is well-based and thoroughly acceptable.

Even in physics, there is no infallible procedure for generating reliable knowledge. The calm order and perfection of well-established theories, accredited by innumerable items of evidence from a thousand different hands, eyes and brains, is not characteristic of the front-line of research, where controversy, conjecture, contradiction and confusion are rife. The physics of undergraduate text-books is 90% true; the contents of the primary research journals of physics is 90% false. The scientific system is as much involved in distilling the former out of the latter as it is in creating and transferring more and more

---

[41] This philosophical model of a science is discussed, with appropriate references, by Mary Hesse (1974) in *The Structure of Scientific Inference* (London: Macmillan). The same metaphor comes naturally to T. S. Kuhn (1962) *The Structure of Scientific Revolutions* (University of Chicago Press), p. 7 'the network of theory through which the [professional scientific community] deals with the world'.

[42] As Bohr said of his theory, 'It should be made clear that this theory is not intended to explain phenomena in the sense in which the word *explanation* has been used in earlier physics. It is intended to combine various phenomena which seem not to be connected and to show they are connected.' W. Heisenberg (1973) in *The Physicist's Conception of Nature* edited by Mehra, p. 264.

bits of data and items of 'information'. In later chapters (4.4, 6.3) we shall observe the growth of a scientific consensus, the way in which it purges itself of errors and misconceptions, and the influence of the 'collective consciousness' of a scientific community on the perceptions of its members. What is surprising is not that each one of us makes many mistakes, but that we have made such remarkable progress together. That is why, when we come to look at the assessment of strong, quasi-logical and mathematical models in the behavioural sciences (7.6) we must not abandon caution and scepticism just because they superficially resemble the historically successful models of physics. For Galileo, Nature was written in mathematical language; but with all his genius he could not read the plain sociological messages from his old friend the Pope.

3
Common observation

'A fool sees not the same tree that a wise man sees'.
Blake

3.1   *Equivalent observers*

The communication system of science determines the *form* of the
messages exchanged between scientists: the *content* of these messages
purports to be information about the 'external' or 'real' world. The
scientific community acquires this information through the bodily
senses of its members. Scientific knowledge is gained primarily by
observation – by 'using one's *eyes*'.[1] Most of the messages that scientists
send to one another are reports of 'what has been seen' under
specified circumstances.

For the moment we may ignore the fact that these circumstances
are often fantastically contrived (3.3) and that the human observation
may be no more than a pointer reading of a complex instrument (3.4).
Nor need we be deceived by the impersonal, passive voice in which
such messages are conventionally cast: 'The litmus paper *was observed*
to turn red'. Everybody knows that this means: '*I* saw it turn red';
or '*We* saw it turn red'; or '*Jones* saw it turn red'; or '*Bosambo, Pooh
Bah* and *Hiawatha* saw it turn red'. The fundamental principle of
scientific observation is that all human beings are interchangeable as
observers. Put *anyone* down in that laboratory, and he or she would
have seen the litmus paper turn red. If this be indeed the case, then
it is a modest logical extension to assert this as a truth for *everyone*;
we are easily tempted to refer to such an observation as an *objective
fact*.

The communal model of science (1.3) restricts scientific information
about the external world to those observations on which independent
observers can agree. It is limited to the *perceptually consensible* as well
as to the *communicably consensual*. The assumption that *all observers are*

[1] Or *ears,* or *nose,* or *taste,* or sense of *touch.* Philosophically and psychologically, these
other senses are not sufficiently different from vision that they need be separately
discussed.

Fig. 3.

equivalent is not merely a basic principle of Einstein's theory of special relativity; it is the foundation stone of all science.

This trite remark conceals many pitfalls, and demands much detailed interpretation and qualification; for the moment let us take it at face value as an unquestioned constitutive principle and see what follows. It strikes deeper at the roots of 'logicality' in science than the positivists seem to realize (5.3).

3.2   *Pattern recognition*

The very possibility of perceptual consensibility depends upon a very ordinary faculty, shared by all human beings and by many animals. Without conscious effort, we all have remarkable skill at *recognizing patterns*.

43

Consider the following familiar phenomenon: two or more people cast their eyes on the same object, or are presented with the same scene; they quickly agree 'That is a such and such'. Try the experiment of showing Fig. 3 to almost any man, woman, or articulate child, and ask them what they see: the answer will be invariably the same 'It's a *man* riding a *bicycle.*' One would be truly astonished if they solemnly affirmed that they saw a *typewriter*, or an *octopus*, or a *potato*. There is almost complete consensus in their reports of what they perceive.

It is true that an occasional observer, with the appropriate life experience, might add, 'The man is Albert Einstein – whom I once knew', or '– whose face is familiar to me from photographs', and might be able to say much more about the background buildings, etc. On the other hand, an Eskimo or an Aborigine who had never before seen a bicycle might not be able to give a satisfactory answer. For the moment, however, let us accept this everyday practical skill in its normal form, without qualifications, and without attempting to analyse the neurophysiological or psychological mechanisms that underlie it (5.1). The point to grasp is that *intersubjective pattern recognition* is a fundamental element in the creation of all scientific knowledge.[2] This is obvious from a few examples.

At its most elementary and primitive level, science includes descriptions and classifications of *natural objects* (plants, minerals, stars, etc.) based upon visual inspection. The practical problem then is how to convey these observations to other natural historians or astronomers. What is the best form of the message to be communicated or stored in the archive? A characteristic example of such a message is as follows:

> 'Deciduous shrub, glabrous or nearly so, with weak, trailing sub-glaucous, often purple-tinted stems, either decumbent and forming low bushes 50–100 cm high, or climbing over other shrubs, rarely more erect and reaching 2 m. Prickles hooked, all±equal. L'flets 2–3 pairs, 1–3.5 cm, ovate or ovate-elliptic, simply, rarely doubly serrate, glabrous on both sides or pubescent on the veins (rarely all over) beneath, rather thin; petiole usually with some stalked glands; stipules narrow, auricles straight. Flowers 1–6, white 3–5 cm diam.; pedicels 2–4 cm with stalked glands, rarely smooth; buds short...etc. etc.'

What is this strange plant? In Fig. 4 we at once recognize a species of rose – in fact the familiar field rose, *Rosa arvensis*. It does indeed have the characteristics listed above; in the picture, however, we

---

[2] This point is emphasized by Polanyi (*Personal Knowledge*, chapter 12) in relation to biology, but applies, in fact, to *all* branches of science.

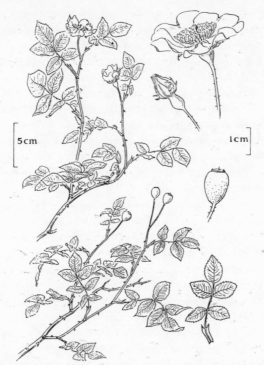

5cm

1cm

Fig. 4. From Clapham, A. R., Tutin, T. G. & Warburg, E. F.
(1960). *Flora of the British Isles, Illustrations II*, p. 15. CUP.

perceive a pattern which the botanist learns to distinguish like the face
of a friend.

Thus, although the verbal description draws attention to important
features which might help us to place this plant in the correct genus,
whilst distinguishing it from other species of rose, it is essentially
incomplete without the picture. The technical words that are used in
this description refer to other remembered visual patterns. How would
one define the adjective 'serrate', except to say that it was 'like a saw'?
Try to imagine a purely verbal account of such an object, without
actually drawing a picture or seeing an actual specimen.[3] We need not

---

[3] As pointed out by N. R. Hanson (1958) *Patterns of Discovery* (Cambridge University
Press) p. 25. 'Pictures and statements differ in logical type, and the steps between visual
pictures and the statements of what is seen are many and intricate. Our visual
consciousness is dominated by pictures; scientific knowledge however is primarily
linguistic.' But Hanson was mainly concerned with theoretical physics, and under-
estimated the place of actual pictorial representations and visual perception in
science as a whole.

The importance of 'thinking with pictures' as an essential strand in technology is
brought out by E. S. Ferguson, 'The mind's eye: non-verbal thought in technology',
Science, **197**, (1977), 827–36.

be led astray by the numbers that occur in such descriptions. These merely summarize, in consensible language, the outcome of other sensory experiences, such as counting and measuring and will seldom be amenable to the logical transformations of any mathematical theory (2.4).

This example was drawn from *taxonomy*, where there is a long tradition of visual recognition and pictorial representation. One of the plants represented in this sixteenth-century Chinese pharmaco-poeia (Fig. 5) is immediately identifiable by a trained Western botanist as *Artemisia alba*. What could demonstrate more aptly the pheno-menon of perceptual consensibility across the frontiers of almost unconnected human cultures?

It is true that this branch of natural history is neglected and scorned by many modern biologists, who prefer to concentrate on the more molecular aspects of the subject. But their research is inevitably para-sitic on a whole body of taxonomic knowledge, accumulated by skilful visual observation and stored as specimens or pictures to be inspected, compared, or recognized from memory. Observation of natural phen-omena, the accumulation of factual data, systematic comparison of specimens and of schemes of classification are essential activities in science, preliminary to, yet never entirely superseded by, theoretical unification at a higher level of abstraction. The ideology of physicalism (2.7) blinds many philosophers of science to this elementary fact about what scientists actually do – and why. Indeed, the extraordinary fact that *consensual* schemes of biological classification can be arrived at by taxonomic research was the enigma to which Charles Darwin devoted his life and triumphantly resolved (7.2).

The recognition of the necessity for precise, critical, intersubjectively validated, visual observation revolutionized *anatomy* and *physiology* in the sixteenth century. In the Middle Ages, the medical sciences were dominated by theoretical systems dating from Antiquity. Occasional pictures, such as Fig. 6 have no direct reference to nature; they are merely diagrams to illustrate the verbal descriptions of the human body inherited from Galen. The transcendental genius of Leonardo da Vinci[4] is nowhere better exemplified than in his anatomical draw-ings, such as Fig. 7. Leonardo was probably the finest scientific observer that ever lived; unfortunately his anatomical work was not

---

[4] Leonardo clearly perceived the inadequacy of verbal description: 'And ye who wish to represent by words the form of man and all the aspects of his membrification, get away from that idea. For the more minutely you describe, the more you will confuse the mind of the reader, and the more you will prevent him from a knowledge of the thing described. And so it is necessary to draw and describe.'

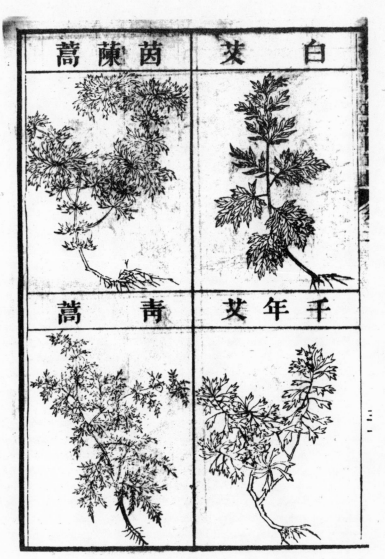

Fig. 5. From Needham, J. (1954). *Science and Civilisation in China*, vol. I, facing p. 164. CUP.

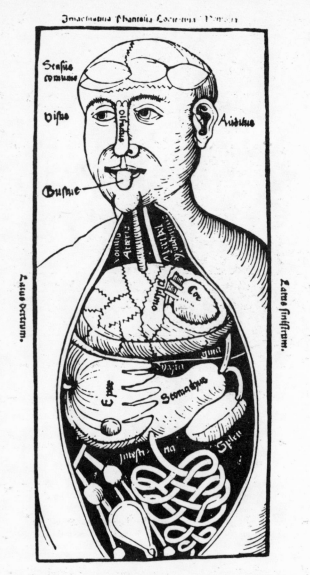

Fig. 6. Diagram of the thoracic and abdominal viscera (1516).

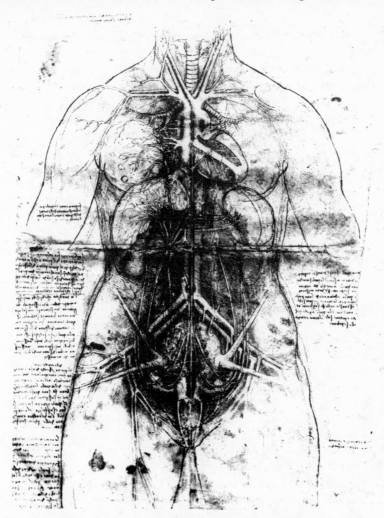

Fig. 7. 'Transparent torso' by Leonardo da Vinci. Photograph from the original at the Royal Library, Windsor Castle (copyright reserved).

published until centuries later.[5] His research programme, which was eventually carried out by Andreas Vesalius, depends absolutely on the consensibility of visual perception. In assessing the 'truth' of such

[5]   This raises a nice question. From the point of view of our 'consensus' model was this unpublished research 'scientific', since it was already obsolete by the time it was eventually made public? We leave such questions to logic choppers, whilst we gaze in admiration at these masterpieces of intellectual power and aesthetic sensibility!

work, the anatomist should, in principle, take little account of theories (4.4); he simply carries out a dissection, and compares the drawings with what he can see with his own eyes. In practice, it is difficult for him not to see what he has learnt to see, under the influence of the accepted paradigm of his subject. But within that limitation of vision, the messages that anatomists communicate to one another, and store as 'objective knowledge' in the scientific archives, are drawings and photographs, to which the accompanying text is merely a commentary.

It would be a great mistake to suppose that the dominant role of visual observation in biology has now been overthrown by more 'analytical' methods. The skilled eye and hand-lens of the old-fashioned 'naturalist' were supplemented in the late seventeenth century by the optical microscope, which has only recently been overtaken in magnifying power by the electron microscope. But the modern cellular biologist is as much concerned to discover and explain the spatial organization, the 'shape', the 'form', the 'structure' of the 'organelles' he can observe within the cell as was the Renaissance anatomist describing the organs he could observe by dissecting a cadaver. A modern scientific paper on *microbiology*, *physiology*, or *pathology* is as heavily loaded with electron micrographs as any mediaeval herbal with its woodcuts of botanical species. The caption of such a figure (Fig. 8) does not read 'Look at this pretty picture!': it tells us to *observe* characteristic shapes which are highly significant and are to be interpreted scientifically. Thus, an immense amount of scientific information is gathered by the direct visual recognition of similarities or analogies between pictorial patterns, mediated solely by the enormous magnifying power of these instruments.

We may also reflect upon the fact that a great deal of experimentation in physiology and *biochemistry* is directed towards the elucidation of 'invisible' spatial structures. In its way, the determination of the 'structure' of a complex molecule by X-ray diffraction is a splendid example of the mathematical transforming power inherent in good clean physics (2.7). But the goal of all the instrumentation, electronics, computation and scientific skill that goes into such an investigation is to produce a three-dimensional model, or a series of pictures, of the arrangement of the atoms in space. This is the piece of scientific knowledge that we shall eventually transmit to our colleagues or put into the archive (4.1) – not simply for 'its own sake', but because we anticipate that visual inspection of the model will suggest entirely new properties of the molecule that would not be immediately

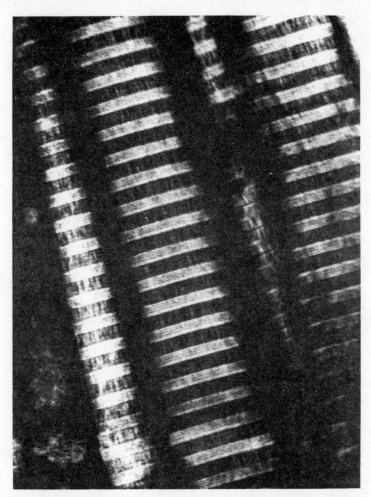

Fig. 8. Electron micrograph of human collagen fibres (mag.
×90000). The reader of the paper in which this picture appeared
was invited to 'note precise axial periodicity'.

detected by mechanical measurements or analytical transformation of
the corresponding data.

What is the goal of modern *chemistry*, but the creation of the perfect
'molecule analyser'. Feed a small sample of an unknown material into
this hypothetical machine: five minutes later, out pops complete
information about the *structure* of its constituents. Ideally, this includes
a photograph of the arrangement of the atoms in each molecule – little

51

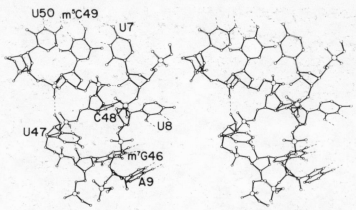

U50 m⁵C49
U7
C48
U47
U8
m⁷G46
A9

Fig. 9. Stereoscopic view of part of a molecule of yeast tRNA[Phe].
From Quigley, E. J. & Rich, A. (1976). Structural domains of
transfer RNA molecules. *Science*, **194**, 796–806. © Copyright 1976
by the American Association for the Advancement of Science.

spheres neatly labelled C, N, O, S, etc. or coloured black for carbon,
white for oxygen, yellow for sulphur, green for chlorine, etc. joined
by neat springy 'bonds', as in one of these delightful sets of molecular
models that no teacher can resist, illustrating a little too concretely the
basic principles of *stereochemistry*, whose imagining revolutionized
chemical theory a century ago (Fig. 9).

Is it necessary to point out that *geology* and *astronomy*[6] are almost
completely dependent on visual observation for their primary data?
As in biology and chemistry, this does not preclude evidence obtained
non-visually by complex apparatus and interpreted analytically by
the application of theory. But it is hard to imagine these sciences
originating and flourishing in the country of the blind.

As an intellectual purist the *physicist* imagines that his science is free
of the taint of observational subjectivity. In principle, the 'data' of
physics are numbers representing pointer readings or computer print-
out from mechanical or electrical apparatus, taken without significant
human intervention. But consider Fig. 10 – not, as it seems at first sight,
camping gear bundled untidily in somebody's attic, but a 'stereoscan'
electron micrograph of the surface of a composite material after
fracture. To careful inspection, this picture reveals the way in which

[6] This is the discipline chosen by D. T. Campbell (1966) in *The Psychology of Egon
Brunswick* edited by K. R. Hammond (New York: Holt, Rinehart & Winston) pp.
81–106, to exemplify the importance of pattern matching in science, in much the same
terms as the present discussion.

Fig. 10.

the reinforcing fibres are pulled from the matrix before they snap, and thus plays an important part in the physical interpretation of a complex but calculable phenomenon. Should such evidence from pattern recognition be denied to the research physicist, on philosophical grounds? Or should we maintain that the quantitative study of the properties of materials is not really 'physics' (2.7) after all?

Visual perception plays a vital part in the most aristocratic branch of experimental science – the physics of elementary particles. One of the main instruments in this field is the bubble chamber, where high-energy charged particles make visible tracks that are easily photographed and inspected for unusual 'events'. Nowadays, to be sure, the eye of the observer is aided by optical and electronic devices, hooked up to a computer. Where the parameters of a large number of similar events are to be determined, these procedures may be almost completely automated. But significant scientific progress in high-energy physics did not have to wait until this approximation to instrumental 'objectivity' had been achieved (5.6). Until quite recently, the fundamental link in the observational chain was a 'scanner' – a

53

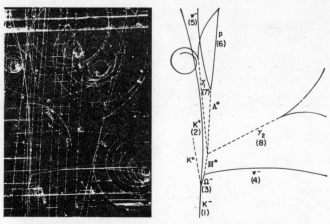

Fig. 11. Bubble-chamber tracks, showing the 'Omega-minus' particle predicted by Gell-Mann and Neeman (see 2.8). From *Physics Today*, April 1964, 57–60. © Copyright of the American Institute of Physics.

person without deep scientific education who was instructed to search bubble-chamber photographs or nuclear emulsions for particle tracks of a particular type (Fig. 11). It is significant that although individual members of a team of scanners might acquire a good deal of practical skill, there was no need for an elaborate 'calibration' of their eyes and brains. This perfectly exemplifies the high degree of visual consensuality that scientists take for granted as they collect information about the 'external world'.

As a final illustration of the scientific application of the maxim that 'seeing is believing', consider Fig. 12. Peculiar markings were observed on the surface of Mars, in the neighbourhood of a great volcanic cone. It was suggested that these might be due to wind-blown deposits of sand. A model 'volcano' was set up in a wind tunnel, and sand deposits allowed to build up around it under the influence of a dimensionally-scaled current of air. The equivalence of the 'natural' and 'artificial' patterns is quite sufficient proof that this explanation must be substantially correct. In other words, this matching of visual patterns has validated a physical theory without recourse to formal mathematical analysis (2.9).

Suppose, indeed, that an attempt were made to verify this hypothesis by actual calculation. The standard aerodynamical equations of motion of the atmosphere would be solved arithmetically in a large computer, with boundary conditions adapted to the supposed shape of the crater

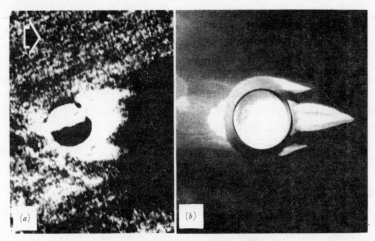

Fig. 12. (a) Photograph of small crater on Mars, to be compared
with (b) distribution of sand around model. From Greeley, R. *et al.*
(1974) 'Wind tunnel studies of Martian aeolian processes'
Proceedings of the Royal Society A, **341**, 331–60.

and the strength of the wind. From the theory of the transport of small
solid particles in a current of air, the applied mathematician would
then compute the distribution of sand deposits in the neighbouring
region. But how would he decide, or demonstrate, that the results of
this computation were in agreement with experiment? Ten to one, he
would instruct the computer to print out a 'picture' of the sand
deposits, *which he would compare visually with the space-probe photograph!*
Here, once more, the pictures themselves constitute the basic 'information' that the scientist conveys to his colleagues.[7]

In this effort to demonstrate the ubiquity of pattern recognition in
science, I have deliberately chosen cases where visual perception plays
different methodological roles (4.1). But whether we are concerned
with biological taxonomy, or with the use of 'graphics' in the 'interactive mode' of an advanced computer technique (5.2), we come back
to the same fundamental point – that the bodily senses are the only
link between a human mind and the world he or she inhabits. Visual
perception, by its intersubjective consensibility, is an essential element

[7] 'A picture is worth a thousand words; a picture is a thousand times less specific than
a short sharp statement. But, by the same token, one word is worth a thousand
pictures; a statement can supply a focus for the attention that is different in type from
anything generable via confrontation with a picture' N. R. Hanson (1970) *The Nature
and Function of Scientific Theories* edited by R. G. Colodny (University of Pittsburgh
Press).

in the creation and validation of scientific knowledge, and pattern matching provides a standard of consensuality which is never completely superseded by more 'objective' devices such as mechanical instrumentation. For the moment, we shall not pursue this theme further; but in Chapter 5 we return to examine the philosophical status, empirical origins and practical capabilities of this familiar yet mysterious faculty.

3.3 *Experiment*

To exploit the principle of the equivalence of observers (3.1) we must ensure that they are looking at the same, or equivalent events. That does not mean that the reports of the individual observer concerning a possibly unique event (such as the landing of a meteorite in his front garden, before his very eyes) are to be altogether distrusted or despised. Where this is the best evidence available, it is worth consideration: the principle has a weaker vicarious interpretation along the lines of 'we may accept as consensible the sort of report that any other human being might equally well have made if placed in the same circumstances'. But the experience and wisdom of the scientific enterprise is that the most reliable 'facts' are those that are attested to by several independent witnesses or by material evidence, such as a photograph (4.3), that has been generated under identical circumstances arranged by independent research workers.

But how to ensure that the circumstances are, indeed, the same? In the practice of science this is not left to chance. Modern science does not depend upon adventitious evidence like the reports of islands collected by ancient geographers from ship's captains blown off course by storms. What we have come to call an *experiment* may have many motives (such as the confirmation of a prediction: 2.8) and its outcome may have many interpretations (2.9), but for our present purposes we should think of it simply as a *specially contrived observation, carried out under controlled, reproducible conditions.*

Therein lies a paradox. 'One cannot step twice into the same river': the very idea of a reproducible experiment implies a firm belief in a certain uniformity of nature (2.8). But this metaphysical principle (6.4) cannot be invoked as a detailed guide to practice. From the very beginning, the achievement of consensus between different observers concerning the results of repetitions of the same experiment is fundamental to the creation of any body of scientific knowledge. It is

comical for us now to read of the early experiments carried out by
the Fellows of the recently founded Royal Society:

> '*Experiments* of destroying *Mites* by several fumes: of the
> equivocal Generation of *Insects*: of feeding a *Carp* in the Air: of
> making insects with Cheese, and Sack: of killing Water-Newts,
> Toads, and Sloworms with several Salts: of killing Frogs, by
> touching their skin, with Vinegar, Pitch, or Mercury: of a Spiders
> not being inchanted by a Circle of *Unicorns-horn*, or Irish Earth,
> laid round about it.'[8]

But their instinct to test every old wives' tale was sound. Until there
was a consensus on such matters, there was no basis for biological
science. And their investigation of 'making Insects with Cheese, and
Sack', was only an early case of research on the 'spontaneous generation of life', where the experimental situation remained chaotic until
Pasteur, in 1860, showed how to prevent dust contamination of a
nutrient broth kept open to the air. The science of bacteriology could
not get under way until research workers could feel confident that
their observations were being performed under these 'controlled,
reproducible conditions'. Pasteur's famous experiment, so convincingly falsifying a widely-held theory, was also the pre-requisite for
consensibility and consensuality in experimental method itself.

In our superior modern way, we tend to think that the marvellous
instruments provided by modern technology will give us 'objective'
experimental results on which we can rely and on which we can
theorize without doubt or controversy. But even the quantitative data
of physics are not won from Nature simply by pressing the buttons
and reading the dials of infallible mechanical devices. Good science
is never that easy!

In his research laboratory, the scientist is always trying to obtain
new results. His experiments cannot be, like those of the teaching
laboratory (6.2), mere repetitions of past triumphs – well-known quantities determined with standard apparatus to a fair degree of accuracy.
No scientific journal will publish data of that sort: who could be
interested in what is already known as a fact?

The ideal type of experimental novelty is one of those dramatic
episodes, by which scientists are inspired, and on which historians and
philosophers of science concentrate their attention, where a previously
unknown phenomenon is observed with a relatively simple technique.

[8] Thomas Sprat (1667) *History of the Royal Society* (1959: Facsimile edition: London:
Routledge & Kegan Paul) p. 223.

It may be an unforeseen *discovery* (3.6) or one of those *crucial experiments* which startlingly confirms a fantastic prediction (2.8) or decisively falsifies a beautiful theory (2.10). In its purest form, it should never have been attempted before, yet should be unambiguous in outcome. An almost perfect opportunity for such experiments was offered to physics in 1956, when Yang and Lee questioned the validity of the 'law of parity conservation' in elementary-particle processes involving the 'weak' interaction. This law, which is so fundamental that if it were not true it would allow us to determine physically whether our universe is 'right-handed' or 'left-handed', had always been taken for granted, and never been experimentally verified. Within a few months, several research groups had devised and carried out various relatively simple experiments, using available apparatus and well-established techniques, which astonishingly confirmed, beyond a shadow of doubt, that parity is *not* conserved in such processes, and that our universe is uniquely assymmetrical in this respect.

Yes, indeed, that is the sort of coup that every scientist dreams of. But the reality is much less exciting – and much more demanding – than the myth. The novelty that can be claimed for most experimental research is that the properties have been measured of a new material or of a process that had not previously been studied in detail, or that an existing experimental quantity has been determined to greater accuracy. It is, one supposes, unromantic 'normal' science, performed conscientiously under the influence of a 'paradigm' (4.4), solving 'puzzles', rather than 'problems'.[9]

The idea of carrying out such an experiment may, indeed, have been discussed widely in that field of research; suitable apparatus may, indeed, be already available; yet the actual performance of the investigation is by no means mere routine like an operation for appendicitis. The fact that it has not been done before is usually no accident; the task to be carried out may be on the very edge of what is judged possible within the 'state of the art'. Difficulties that have vitiated previous work must be avoided; the apparatus may have to be driven to the limits of its performance; the skill and experience of the experimenter are put to the test as they are in all high arts.

Nowhere is the pursuit of excellence more demanding than in experimental research. There is intense individual motivation to get the *best* results possible, in terms of accuracy and theoretical significance. To understand this, we must think back to our model of

[9] This nomenclature is due, of course, to T. S. Kuhn, *The Structure of Scientific Revolutions*.

the scientific community (1.3). The dynamic of research does not derive from an abstract metaphysics, nor from conscious adherence to a normative code; it is driven by the psycho-social tensions between critical and creative roles, between competition and coöperation (6.3). There is no *formal* machinery within the scientific community for inspecting the record of reported experimental results and assigning specific tasks of verification in doubtful cases (6.5). This should happen 'automatically', by the independent decisions of several research workers to repeat 'for themselves', an interesting experiment – not simply to be quite happy with the *same* answer, but to *improve* on the previous results of others. Under the lash of scholarly competition, they unconsciously act out the Popper principle of falsification (2.9), not only in relation to the theories but even against the objective 'facts' claimed by their rivals. In this process they themselves make evident the hidden deficiencies of the supposedly infallible mechanical devices on which at each stage we had all hoped to rely!

This is a point which the layman or philosopher often ignores, but which is all too familiar to the practising scientist. In the long run, the observational and experimental data of science are well tested by a variety of independent investigations. But results whose very novelty makes them particularly interesting (3.6) often turn out later to be wildly inaccurate because they are being produced at the very limits of experimental technique. Further experiments designed to check or improve the data are not necessarily better conceived, and may be equally unreliable. In some cases, where the effect to be measured is small, or where the conditions of a specimen are very difficult to control[10] the experimental situation may remain chaotic and controversial for many years, despite every effort to 'solve' this little 'puzzle' of 'normal' science.

Observational consensibility and consensuality cannot, therefore, be taken for granted in science. They must be sought deliberately, by artful experiment, by clever technique and by all the resources of critical controversy. Looking back over a long history, we may be

[10] I have in mind, for example, the determination of the *Hall effect* which is of great interest in the theory of liquid metals. An electric current is passed through a liquid metal (at a temperature of, say, 1000 °C) in a strong transverse magnetic field. A very small voltage has then to be measured across the specimen. Every schoolboy physicist can imagine the problems of containing the sample, maintaining it at a uniform temperature, avoiding convection currents and spurious magnetohydrodynamic voltages, damping down vibrations, amplifying the signal, etc. etc. It is scarcely surprising that a decade of experimental work by several very accomplished research workers has not produced an agreed set of data for this (in principle) basic and elementary physical parameter.

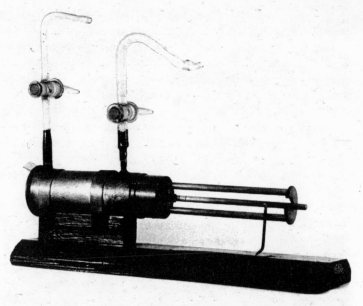

Fig. 13. The apparatus with which Ernest Rutherford 'split the atom'.

astonished at the skill and precision that are now commonplace in what was once a very shaky business; but the questions we ask grow sophisticated in parallel with the means available to answer them, and the experimental evidence deemed relevant to the scientific problems of today often turns out to be just as questionable, at its own level, as were some of those reports on which the Fellows of the Royal Society were seeking reliable information 300 years ago.

3.4 *Instrumentation*

Science, in our time, has become an *industry*. Research is no longer an activity of individuals, but is carried out in *teams*, where the labour is subdivided and specialized. Research apparatus is so complex and expensive, and is capitalized on such a large scale, that it far transcends the intellectual or manipulative abilities of a single research worker to design, construct, or operate on his own. This historical development is of great significance in the sociology of the scientific community and in its relations with society at large; it also has some influence on the contents and authority of scientific knowledge.

Consider, for example, the famous experiments of Ernest Ruther-

Fig. 14. A modern machine for 'splitting the atom'. Foto: A. Zschau. GBI. Gesellschaft fur Schwerionenforschung (GSI), Darmstadt. Luftaufnahme freigegeben unter Nr. durch den Regierungsprasidenten in Darmstadt.

ford and his students on radioactivity and the structure of the atom. The prototype apparatus for all particle scattering experiments (Fig. 13) was so simple that it could be made by a skilled craftsman with his own hands, and thus so cheap that it could have been bought for a few pounds by an aspiring scientist. The technique of the experiment – counting the scintillations produced by alpha-particles hitting a screen after scattering from the atoms in a thin foil – was extremely laborious, but required no special skill, and the mathematical analysis of the results was well within the grasp of any student of physics. In other words, here is a genuinely reproducible experimental situation, of the type envisaged in our model of the scientific process, whose set-up and outcome are readily consensible within an elementary physical framework.

That was half a century ago. Contrast this with a modern piece of apparatus for essentially the same type of experiment (Fig. 14). Here also, we study the scattering of a beam of energetic particles by the nuclei in a solid target. Apart from a factor of a thousand or so in the energy of the beam, it is the same sort of physics, and would be represented theoretically by the same geometrical diagrams and the same equations. Yet this instrument is so large that it occupies a site and buildings that would accommodate a steel mill. It is so complex that its design, construction and maintenance demand the labour of physicists, engineers and skilled workmen by the hundred. The expense of building and running it is a special item in the national budget of a great industrial nation. Each experiment must be planned years in advance, and involves the collaboration of dozens of scientists and technicians in highly specialized roles. And even the intellectually significant task of analysing the results for communication to other scientists has to be handed over to a computer.

The Gargantuan scale of such an instrument has two consequences for scientific epistemology. In the first place, the physical complexity of the apparatus, involving the harmonious interaction of many separate elements – beam magnets, vacuum systems, accelerating voltages, target assemblies, spark chambers, pulse-height analysers, etc., etc. – demands elaborate rationality of design, beyond the grasp of any one person. The results of the experiment are irretrievably embedded in the design theory of the system and all its parts, whose correct working must be taken for granted as the controlled background of the observation. In the end, we cannot say whether the data are derived primarily from the 'external world' or from the theories they are supposed to be validating or falsifying. By their immense practical experience and consummate technical skill, the 'experimentalists' do achieve remarkable and essentially reliable results, which are undoubtedly of significance in building up physical theories. But the circumstances thus contrived are so bizarre,[11] and the recorded data are so recondite, that it becomes difficult to believe that they have any application outside of those very circumstances. Without seriously doubting the validity of the interpretations that are proposed for these phenomena, one may be forgiven for wondering whether this type of research has now been completely captured by the imperatives of its own observational technique. Mutually reinforcing paradigms of experiment and of successful theory (4.4, 6.4) are producing an 'island universe' of scientific knowledge of extraordinary internal subtlety and beauty, but essentially beyond human reach without long years

[11] As has been said of some experiments in high-energy physics: the process to be observed has never occurred before in the history of the Universe; God himself is waiting to see what will happen!

of training, and almost entirely irrelevant to affairs here on earth. In this respect, high-energy physics, as the manifestation of a century's effort to carry the programme of *physics* (2.7) to its ultimate end, is uniquely vulnerable to a sceptical whisper: 'So What?' But the same weakness is risked wherever the instrumentation of research exceeds the power of the human intellect to comprehend its inner working.[12]

The other epistemological effect of the trend towards Big Science is in the reproducibility of experimental results, which is their ultimate guarantee of consensuality and reliability. The psycho-social system of science is assumed to act within a framework of individual incentives and capabilities, where quality control is maintained by competitive bargaining of 'buyers' and 'sellers' of information. The assumption is that this is a 'peasant market', where the dissatisfied purchaser can turn to an alternative producer, or even set up himself to make shoes or raise pigs. In other words, there is normally intense competition in the *quality* of the experimental results of independent research workers (3.3). To pursue the market analogy, the scientific productivity of a 'Big Machine' is enormous, but the mechanisms of quality control may be inadequate because of the immense expense of competitive reproduction of the experiment in another place. The number of such instruments in the world is very limited, and each is heavily committed to what is hoped to be an interesting scientific programme with novel results. There is thus a tendency to avoid what is called 'wasteful duplication of research', and many experiments must wait long before they can be confirmed by independent repetition.

At the present time, the habit of harsh criticism and competition between the leading research teams (6.3) is probably sufficient to prevent this sort of degradation of experimental data. But the movement in high-energy physics towards the construction of the VBA ('the Very Big Accelerator') to serve the scientists of the whole world could, in the end, be damaging to the health of this branch of science. What system of competitive independence for research teams will be

---

[12] We touch here upon the problem of the status of theoretical entities (5.4). In the present context there is much to learn from the evocative remarks of G. H. Mead (1938) in *The Philosophy of the Act* (University of Chicago Press) p. 32. 'The ultimate touchstone of reality is a piece of experience found in an unanalyzed world. The approach to the crucial experiment may be a piece of torturing analysis, in which things are physically and mentally torn to shreds, so that we seem to be viewing the dissected tissues of objects in ghostly dance before us, but the actual objects in the experimental experience are the common things of which we say that seeing is believing, and of whose reality we convince ourselves by handling. We extravagantly advertise the photograph of the path of an electron, but in fact we could never have given as much reality to the electrical particle as does now inhabit it, if the photograph had been of aught else than glistening water vapour.'

built into the bureaucracy that must eventually allocate research opportunities with this instrument? Could one feel quite sure that the results of incompetent research will be replicated, and errors eliminated, within such an organisation?

From the traditional viewpoint of philosophy, such mundane sociological considerations have nothing to do with the 'problem of scientific epistemology'; scientific knowledge should be able to stand up for and justify itself from its own cognitive resources or metaphysical foundations. But worldly wisdom teaches us to be a little suspicious of knowledge that emanates from a unique, compactly organized and self-regarding social institution, staffed by specially trained and thoroughly indoctrinated personnel (6.2) using machinery of awesome size and complexity. The fantasy of an intellectual conspiracy, entertained by some ideologists of the sociology of knowledge (6.4), has no substance within the natural sciences of today, but one might fear such a corruption from the monopolistic monasticism of the Very Big Science projected for tomorrow.

3.5 *Signal or noise?*

It is a commonplace of elementary scientific method that every experimental result is subject to some degree of uncertainty. In the teaching laboratory we instruct the student to determine the relationship between the pressure and volume of a gas, and take pleasure in his educational experience that the points do not all lie precisely on the hyperbolic curve of Boyle's Law. The logic of the empirical (2.6) is made three-valued by *experimental error*. The aim of good research is to reduce such errors to a minimum; but they can never be entirely eliminated. Part of the practical lore of research is the technique of estimating the magnitude of the uncertainties, in advance as the apparatus is being designed, and retrospectively as a characteristic of the published results. A properly phrased scientific communication should never be a categorical assertion, but should always convey the author's assessment of the credibility of his own claims.

It is easy for the armchair epistemologist to argue that every experimental observation should be repeated, and the apparatus refined, until the data are of the irrefutable accuracy required to confirm or falsify theory. But this ideal may not be a practical goal. It must be emphasized that *statistical* uncertainties are intrinsic to many types of physical experimentation.

Again, elementary-particle physics exemplifies the issue. Suppose

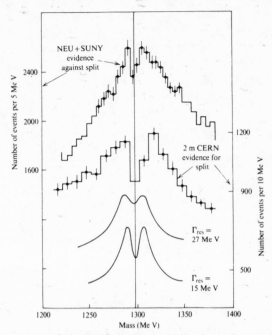

Fig. 15. From B. Maglich. A$_2$ meson debate, letter. *Physics Today*, February 1972.

that we are looking for evidence of the existence of a very unusual particle, in the form of a very rare 'event' on a bubble chamber photograph. It is a fundamental principle of statistics that a phenomenon that occurs, say, once in a million trials *on the average* will not be observed regularly once *every* million trials. The results of any experiment that runs for a finite time (e.g. to collect and analyse ten million photographs) are thus subject to significant statistical fluctuations, and can never be precisely reproduced from one experimental run to the next. In the case I have quoted, for example, there is a finite probability that the rare event may not occur at all, in all those photographs, so that the experiment would suggest (falsely) that the particle does not exist.

It is all very well to insist that we should go on taking pictures until the case is proved 'beyond peradventure' (whatever that means!). Such perfection may cost enormously in money, and in time. The aim of the research is to produce a publishable scientific result, of adequate plausibility, not complete proof. Consider the story of the A$_2$ meson

65

(Fig. 15). If this 'particle' exists, then the histogram should have a double peak. One set of data, obtained at CERN, were claimed to favour this hypothesis; the other results, from Brookhaven, contained rather more events, and were said to be unfavourable to this interpretation. Which are we to believe? It is doubtful whether any principle of statistical inference could yield a foolproof formula that would, say, falsify the hypothesis of 'splitting', to a high degree of probability, on these data. Even the most sophisticated and expensive instruments do not always feed completely certain 'facts' into the scientific information network and its archives.

There is no need, here, to burrow down below metaphysical ground level, to the Heisenberg uncertainty principle of quantum theory. Real apparatus is macroscopic, and seldom pretends to control the parameters of the microscopic systems that are being studied. We have to take what nature offers us, inchoate streams of atoms, in imperfect geometrical configurations, in the limitations of detection and amplification of small signals.

In the physical sciences, experimental uncertainties can take on an active role. To improve the response of our apparatus to small input signals, we turn up the gain control; high amplification makes it sensitive to a variety of other unwanted and irrelevant influences – stray electric or magnetic fields from other parts of the equipment, vibrations of the building, cosmic rays, temperature variations, etc. In response to such stimuli, the pointer of our recording galvanometer is continually aflutter; the actual effect that we are trying to measure may seem no more than a slightly larger wiggle on the output trace. In other words, the scientific 'message' can scarcely be heard against a background of *noise*. In very sensitive instruments the unavoidable uncertainties and errors of observation are magnified into an apparently autonomous random disturbance, impishly impeding the honest search for truth.

The techniques used to separate the desired signal from the noise are very important in highly instrumented science, and cannot be ignored in any assessment of the ultimate reliability of scientific knowledge. Methods of *data processing* and *image enhancement* that were developed to improve the transfer of information from unmanned space-probes can now be applied to optimize and automate what were once the prized skills of professional observers. Fantastically accurate measurements are achieved by proper averaging, spectral analysis, aperture synthesis and other computational devices. It almost seems as if the intrinsic limitations of observation can always be transcended by such tricks.

But these techniques may not be so efficient when we look for a signal that may not be there at all, or whose basic characteristics can only be guessed. Consider the search for *gravitational waves*. *In principle* – i.e. by mathematical deduction from Einstein's equations of general relativity – the rapid acceleration of a massive object generates a wave-like disturbance in the space–time continuum. *In principle*, this perturbation of the gravitational field would induce mechanical strains in any rigid system, which could be detected as vibrations or small changes of length. The theory is now well understood and says pretty clearly that no *known* astronomical process, such as the collapse of a star, would produce anything like sufficient gravitational radiation to be detected by any mechanical system that could at present be built on earth. So why try?

Nevertheless, back in the early 1960s, Joseph Weber of the University of Maryland set up a 'gravitational antenna' – a 1½ ton cylinder of aluminium, delicately suspended, and isolated from all other influences save those of gravitation. The vibrations of this cylinder were then recorded by instruments so sensitive that they could detect length variations of the order of one-millionth of the diameter of a single atom. Naturally, the observed output of this apparatus is fluctuating 'noise' due to innumerable small disturbances that could not be completely screened out. Thus, the question whether any particular wiggle might be due to the passing of a gravitational wave through the antenna cannot be decided by mere inspection of the record, for we have not the slightest idea what the form of the wave should be, nor when it might be expected to arrive.

Much to everybody's surprise, however, in 1969 Weber reported a number of 'events' corresponding to coincidental disturbances at two such antennae, one in Maryland, the other in Chicago, 1000 km away. It seemed very unlikely that the same extraneous force could be acting simultaneously at such distant points unless it was, indeed, of extra-terrestrial origin. If electromagnetic disturbances, such as the radiation from solar flares, could be excluded by good instrumental design, then this must be regarded as serious evidence for gravitational waves of much larger amplitude than could possibly be accommodated by current astronomical theory.[13]

The world of physics was amazed, but not turned upside down by this extraordinary discovery. The design of the apparatus, and the techniques of data reduction were subjected to highly critical scrutiny. Controversy arose over the statistical analysis of the results, and not

[13] For details of this episode, see 'Gravitational Waves – Progress Report' by J. L. Logan in *Physics Today*, March 1973, 44.

all doubts were satisfactorily resolved. The cosmologists looked hard at the theory, and allowed certain hypothetical possibilities for unusual astrophysical events that might be taking place in the obscure region at the centre of the galaxy. The experimental evidence, if it is correct, must take precedence over speculative theory!

But other groups of physicists, in many countries, constructed their own gravitational antennae, and looked for similar signals. These instruments, designed to equal or surpass Weber's apparatus in sensitivity, have *not* confirmed his observations. Thus, although the 'events' reported by Weber have not been satisfactorily explained away, there is now general agreement that the gravitational radiation reaching the earth from outer space is below the present levels of physical detection. Much more sensitive instruments are being designed; the research continues; but Weber's observations do not form part of the body of agreed 'facts' about the world of nature that must be taken into account in a general theoretical synthesis.

This episode shows the 'scientific method' at its best (6.3). It is interesting to note, however, that there was no attempt to replicate Weber's apparatus in every detail.[14] The various 'gravitational antennae' that were set up were of a variety of shapes and sizes, and many entirely different physical principles were exploited to detect the strains induced by possible gravitational disturbances. In other words, the nature of the phenomenon was supposed to be perfectly well understood within the framework of our current knowledge of physics (quite apart from the astrophysical mechanisms by which the radiation might be generated), and the best possible techniques were devised, by each group, to detect it experimentally. Nobody supposes, for example, that there is some mysterious, accidentally efficacious quality of the Weber instrument that makes it susceptible to an influence that is otherwise beyond the reach of our observation or understanding. Thus even an extraordinary scientific discovery, almost as inexplicable to the knowledgeable physicist as the sight of an unidentified flying object (6.6), has to be treated within the general scientific culture of our time, which it could only transform if it were confirmed, again and again, by many independent observers, and made much sharper, much more easily reproducible, much more circumstantial and detailed, by subsequent research.

In the physical sciences, we can often reduce the level of background noise by improvements in our apparatus. But the intrinsic variability

[14] H. M. Collins (1975) *Sociology*, **9**, 205–24.

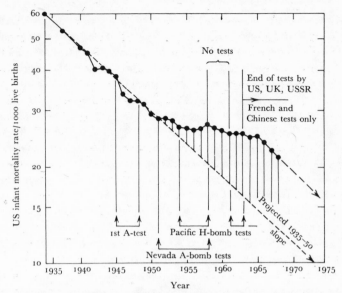

Fig. 16. After Sternglass, E. J. Infant mortality and nuclear tests. *Bulletin of the Atomic Scientists*, April 1969, p. 18.

of biological organisms and social institutions (7.2) is usually quite incorrigible, and the design of a deliberately causal experiment (7.4) runs into ethical or political obstacles – or ends up by killing the goose that laid the golden eggs. In desperation we invent refined techniques of statistical analysis to separate the 'signal' from the 'noise', but the uncertainties of strict causality may turn out to be irreducible (7.5).

Thus, the counterpart of Weber's experiment might be the data collected by Ernest Sternglass (Fig. 16) to show a possible influence of atmospheric nuclear tests on infant mortality. Inspection of these data certainly suggests such a connection whose mechanism could easily be imagined as a biological consequence of enhanced radioactivity in the environment. Yet one sees at once the difficulty of confirming or falsifying this hypothesis. It is not certain, for example, whether the numerical quantity 'infant mortality rate' has an invariant definition or significance over a number of years, or what other causes might have been at work to alter it.[15] There is no strong framework of precise and rigorous theory in which this hypothesis would have a definite place, with other definite, perhaps observable consequences.

[15] See e.g. E. J. Sternglass (1969) *Bulletin of the Atomic Scientist*, April 1969, 18; P. M. Boffey (1969) *Science*, **166**, 195; and many other comments at the time.

Since a direct experiment is clearly out of the question, we can only regard such observations as suggestive and worth recording, but very far from providing reliable evidence for, or against, any such effect. Epidemiology has immense practical value in alerting society to hidden dangers, or making evident some unsuspected association that might have possible causal origins, but it is a crude technique of research, and there is no 'instrument' of statistical analysis that will filter the desired signal from all the noisy data with which it has been scrambled. The implied analogy between a sophisticated technique of mathematical interpretation and a piece of complicated physical apparatus is facile and misleading. In the behavioural sciences, the meaning of the 'signal' itself, even if it can be clearly observed, is often in doubt, and cannot be supposed intelligible just because, in some trivial sense, it happens to be 'there'.

The point to be emphasized is that the information exchanged between scientific observers, and eventually transmitted to the archive as consensual elements of public knowledge (4.2) does not include every pointer reading of every physical instrument, nor the answers given to every sociological questionnaire. The raw data must be refined, processed, analysed and interpreted by each research worker in his own laboratory, before they can be made sufficiently compact and sufficiently interesting, for onward transmission. These processes are, themselves, heavily laden with theory, and deeply embedded in the current scheme of thought (4.4). It is the task of the individual observer to minimize, but not to underrate, the noise content of his data, for it will be the goal of the scientific community to select the correct signals from the background rubbish, and to give them out, stripped of all apparent uncertainty, as reliable knowledge and the scientific truth.

3.6 *Discovery*

To most people, science is a process of *searching*, in which the facts of Nature are *discovered*. This definition draws attention to the ever-widening scope and novel contents of scientific knowledge. Although (1.3) we deliberately avoid discussion of the psychology of research, and adopt a relatively short-term view of the state of knowledge at a given moment, we cannot fail to notice that a special status is often given – by laymen, as well as by scientists – to what are described as 'important scientific discoveries'.

In a very general way, all the well authenticated results of research might be called 'discoveries'.[16] But the word has a more specific reference, to the outcome (often accidental) of a particular kind of experiment or observation. What it really means is: the falsification of an assumed null hypothesis. We say that Christopher Columbus *discovered* America because he falsified the hypothesis – previously unquestioned or taken for granted – that there was nothing but ocean between Europe and China.

What makes a discovery important and scientifically exciting is the degree of surprise that it occasions. The discovery of the coelacanth was remarkable because it falsified the 'lazy' hypothesis that creatures of this kind, well known from very ancient fossil records, must have long been extinct. One of the most important astronomical observations of recent years was the discovery of *pulsars*. Signals repeated at very precise intervals were recorded in the 'noisy' output (3.5) of a large radio telescope. Anthony Hewish and Jocelyn Bell were so astonished by this observation that they had to assure themselves by very careful experiments that this signal really was coming from outer space – that it wasn't just the ticking of somebody's electric clock. It was scientifically of extraordinary interest because it falsified one of the standard hypotheses of astrophysics – that the radio-emissions of astronomical systems such as stars, being due to random natural processes, are always irregular in time and amplitude.

This characterization suggests that even discovery, which we think of as novel in content, a 'step into the unknown', is deeply connected with current models of reality (2.5, 4.4).[17] It is purely a question of definition whether we describe the successful search for a theoretically predicted phenomenon, such as the Omega-minus particle (2.8), as a discovery in the fullest sense or whether it merely counts as the convincing confirmation of a plausible theory. The experiments demonstrating the fundamental lack of mirror symmetry in certain types of nuclear process (3.3) could be described either as the confirmation of the hypothesis of Yang and Lee, or, as the falsification of the widely assumed principle of parity conservation. In other words, the concept of a discovery has meaning only in relation to the knowledge available to the discoverer at the time, and does not signify an entirely distinct category within the established contents of science.

[16] This must be the meaning implied by the so-called '*Discovery Method*' in the teaching of science. More accurately, and less tendentiously, this should be called the *Exploratory Method* to indicate the supposedly uninstructed and unprejudiced process of aimless investigation that the hapless pupil is encouraged to undertake before, hey presto, he hits upon another great, already well-known truth!

[17] In science, to echo Beethoven's dictum about music, 'Everything should be both surprising and expected'.

Nevertheless, any recent scientific discovery constitutes an important salient in the front-line of research. It draws much attention to itself, and demands a sharper assessment of its reliability than more mundane, well-established 'facts' (6.5). By penetrating into unknown territory, it seems to have wide but disturbingly uncertain consequences for previously settled opinions, with implications for revolutionary change. The great prizes of science go to the successful discoverers, not to those who follow behind and carefully explore the new land that has been opened up.

The prospect of fame tempts a few scientists to make inflated claims for their observations. The reward for an important discovery generates pathologies such as the supposed confirmation of a prediction on the basis of quite inadequate evidence. We have seen the way in which Weber's gravity wave observations eventually failed to win credence (3.5). But the search for hypothetical entities is an enduring theme in experimental science.

Thus, for some years theoretical physicists have been saying that the so called 'elementary particles' are in fact combinations of yet 'more elementary' objects called *quarks*, which ought, presumably, to be observable in particle reactions at very high energies. Since the electric charge of a quark is supposed to be somewhat less than that of an electron or proton, it should be fairly easily recognizable. Sure enough, some weakly ionizing tracks of cosmic rays in a cloud chamber photograph (Fig. 17) were given this interpretation.[18] If this evidence were indeed to have proved genuine – that is, if other competent scientists had been won over by this observation, or if they had found similar evidence in independent experiments – then this would have been a famous discovery. In fact, however, it was easy to suggest much more conventional explanations of the weakness of the tracks, and the free quark remains undiscovered – if not unsung. It is not to the credit of the scientific community that very similar claims to have observed *magnetic monopoles*[19] were widely publicized in the mass media before expert re-assessment had demonstrated the weakness of the evidence. These episodes illustrate the necessity for scepticism (5.6, 6.3) and caution in accepting every new and exciting scientific discovery as a genuine revelation.

A more elaborate and instructive episode was the discovery of *anomalous water*.[20] In the early 1960s, a distinguished Russian physical

[18] See e.g. *Physics Today*, October 1969, 55; or *Science*, 26 September 1969, 1340.
[19] See e.g. *Science*, 10 October 1975, 137; or *Physics Today*, October 1975.
[20] For a readable account, see the article by Leland Allen in *New Scientist*, 16 August 1973, 376.

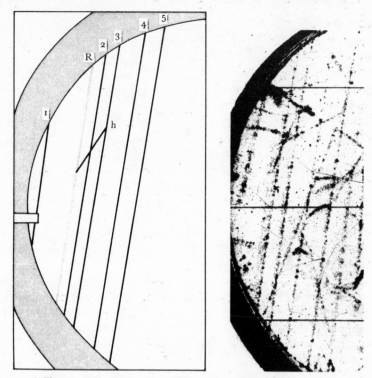

Fig. 17. From McCusker, C. B. A. & Cairns, I. *Physics Today*, October 1969, p. 55. The track marked R was attributed to a quark.

chemist, Boris Derjaguin, reported strange phenomena in water condensed from the vapour in fine glass capillaries. The high viscosity, strange behaviour on melting, unusual Raman spectrum, and other properties suggested that this might be a new phase of water: in other words, it was a remarkable discovery, falsifying the 'lazy' hypothesis that the compound H_2O has only one liquid form.

When this work became known in the West, it triggered off an explosion of research effort (Fig. 18). The material could only be made in small quantities, but further experimental observations catalyzed a variety of speculations concerning its probable molecular structure. Thus, it was soon argued that it must be a polymer of ordinary water (Fig. 19) and hence given the correct technical name *polywater*. Psychologically speaking, structural models like these, although in fact purely conjectural, give an air of reality and concreteness to the whole subject (4.4). But there was some public scepticism. Joel Hildebrand

73

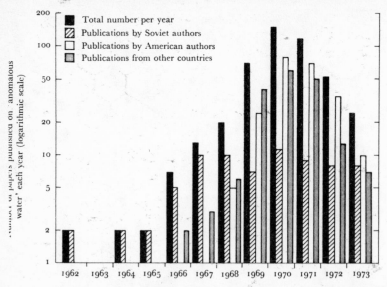

Fig. 18. Papers published on 'anomalous water' each year. After Gingold, M. P. L'eau anomale: histoire d'un artefact. *La Recherche*, April 1974, **5**, 390–3.

wrote sarcastically that he found the product 'hard to swallow' and that he 'choked on the explanation that glass can catalyze water into a more stable phase'.[21] In reply, Leland Allen claimed that 'It almost seems like a model for the ideal research topic that one always tells one's bright young students to be looking for'.[22]

Within a few years, however (and following the publication of several hundred scientific papers, representing several million dollars/ pounds/roubles, etc. of research effort) it became clear enough that the anomalous properties of the condensate were not those of pure water, but must be due to a variety of chemical impurities dissolved from the glass.[23] The discovery was invalidated, and the foundations of thermodynamics stood firm. Whether or not one accepts Leland Allen's retrospective judgement that 'The polywater phenomenon has further verified the efficacy of the scientific method and the always difficult requirement of seeing a problem through to completion in

[21] *Science*, **168**, 1397 (1970).
[22] *Science*, **169**, 718 (1970).
[23] One contaminant was identified spectroscopically as human sweat, from whence came the wisecrack that 'anomalous water was like the classic definition of genius: 10% inspiration and 90% perspiration'.

74

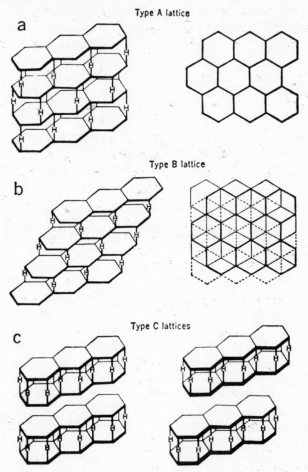

Fig. 19. 'Three types of anomalous water lattices'. From Allen,
L. C. & Kallman, P. A. (1970). *Science*, **167**, 1443. © Copyright 1970
by the American Association for the Advancement of Science.

the face of overwhelmingly negative bias from one's peers', it was
undoubtedly a most instructive episode for the sociology of modern
science.

But our main point here is to notice the uncertainties and errors
that can accompany good experimental technique by highly competent
research workers, and the necessity of independent replication and
verification of the results of observation if we are to acquire reliable

empirical knowledge concerning the external world. We see once more that the credibility of scientific knowledge does not depend ultimately on its having been acquired by the mechanical operations of complicated instruments but on the structure of social relationships that link scientific observers and benevolently rule their inimitable individual powers of perception and cogitation.

4
World maps and pictures

'A mighty maze! but not without a plan.'
                                    Pope

4.1 *Material maps*

Scientific knowledge is not just the sum total of all the observations, 'facts', pointer readings, data, numbers, photomicrographs etc., accumulated by reliable scientists and their instruments. Such an aggregate, however consensible, reproducible, unfalsified, etc. in detail, would be completely incomprehensible without over-riding ordering principles that effectively summarize its content. To those who seek it, science conveys its own interpretations in the form of *theories*. It is the duty of those who participate in scientific activity to generate, communicate and publicly assess the theory of their subject with the same earnestness of purpose as they devote to obtaining reliable consensible, consensual experimental results. Although 'theorizing' may refer only at second hand to the contents of primary observational messages, it is part of the cognitive content of science, and subject to all the imperatives of consensibility and consensuality. Indeed the basic epistemological question – is scientific knowledge *reliable?* – is addressed much more to the theoretical interpretations of research than to the simple hard facts.

In the physical sciences, theories are usually articulated around *models* (2.5) whose properties are capable of relatively precise mathematical analysis. But the concept of a physical model is somewhat too strong and definite as a synonym for theory in most branches of science. The basic conditions for observational consensibility do not absolutely demand geometrical precision and quantitative measurement; they can be met quite adequately by mutual recognition of significant patterns (3.2). In the same way, adequate consensus may be achieved in the realm of theory – i.e. in the generalized, abstracted representation of a body of detailed observational information – in the form of a 'pattern' which may be accepted, recognized and assimilated intellectually without necessarily being capable of complete definition and analysis in formal mathematical or logical language.

It is natural to refer to such a representation as a *map*. It is important to emphasize that this reference is itself metaphorical. Scientific knowledge is a peculiar epi-phenomenon of human existence, and can only be uniquely itself. There seems no absolute necessity that it should be structurally isomorphous with anything so topologically specialized as, say, a graph of vertices (in map language, 'places') connected by edges (e.g. 'roads') on a manifold ('sheet of paper') of a few (two, or perhaps three) dimensions.[1] It could conceivably – and perhaps at times ought (6.7) – to take wilder, more diffuse forms.

But the metaphor is extraordinarily powerful and suggestive.[2] There are good reasons to believe (5.1) that human beings are adapted neurologically and psychologically to comprehend information presented in map form. They can thus scarcely avoid giving this form to the ordering principles they apply to the knowledge acquired in scientific research. The intellectual processes associated with 'reading' a map come as easily and naturally to us as those connected with speech; any alternative representation would seem perverse and incapable of consensuality. Here, of course, we follow Kantian principles in supposing that the form of our knowledge of nature is determined by the structure and capabilities of the human mind.

The communication of scientific observations in the form of consensible patterns (3.2) generates a good deal of *information* that is received, stored and readily made available as useful *knowledge* in the form of *material maps* – two- or three-dimensional representations of the spatial arrangements of material objects. Thus, a great part of 'what is known' to the science of *geology* is precisely what is to be found on a map like Fig. 20. Of course, the data from which this map was drawn could have been stored in the memory of a computer, from which it could be abstracted, transformed and analysed to answer particular questions such as the quantities of rock of various types that

[1] This quasi-mathematical description brings out the *network* characteristics of scientific theory, emphasized (2.10) by Mary Hesse, *The Structure of Scientific Inference*. But, as we shall see, the map metaphor is somewhat broader in scope.

[2] So much so that it flows naturally from the pen of any thoughtful commentator on the philosophy of science: for example Michael Polanyi *Personal Knowledge*, p. 4: 'all theory may be regarded as a kind of map extended over space and time' or Thomas Kuhn, *The Structure of Scientific Revolutions*, p. 108: '[scientific theory] provides a map whose details are elucidated by mature scientists'. The analogy is discussed at some length in Chapter 4 of *The Philosophy of Science* by Stephen Toulmin (1953: London: Hutchinson) but without making the essential reference to the consensibility of pattern recognition. W. H. George (1938), in *The Scientist in Action* (London: Scientific Book Club) takes the same point of view.

Fig. 20. One of the earliest geological maps, by William Smith, 1815.
From *A Delineation of the Strata of England and Wales...*, 1815.
Detail of plate XI.

would need to be transported in the construction of a motorway. But
all the essential information about the topographical relations of the
various mineral species is in the map itself, and the processes of real
interest to geologists can scarcely be grasped except in mappable
form.[3]

Again, the basic chemical goal of determining molecular structures
is directed towards creating an archive of *chemical knowledge* in the form
of diagrammatic 'maps' such as Fig. 21. In the early days of chemistry,
a 'formula' for the relative proportions of the constituents (e.g. H_2SO_4)
was deemed to convey sufficient information about a particular compound: it is now clear that something approaching a stereochemical
model of the configurations of the atoms in space is needed to convey

---

[3] Indeed, it can be argued that the appearance of the new science of geology around
1800 was dependent upon the invention of a 'language' by which its subject matter
could be communicated. 'The emergence of a visual language for geological science'
(M. J. S. Rudwick 1976: *History of Science*, **14**, 149–95) '...helps in a small way to
counter the common but intellectually arrogant assumption that visual modes of
communication are either a sop to the less intelligent or a way of pandering to a
generation soaked in television'.

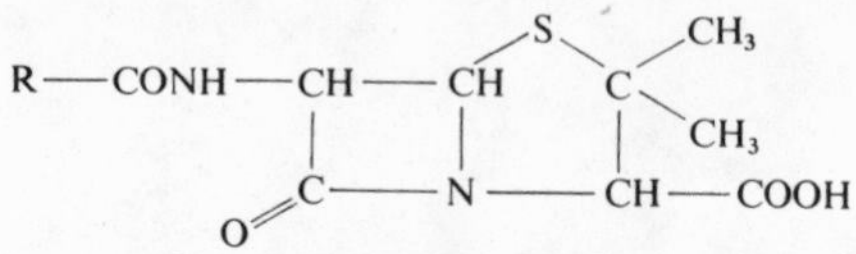

Fig. 21. The structure of penicillin.

'what is worth knowing' about each molecule (3.2). Long and expensive research effort was expended to arrive at Fig. 21 which represents, in diagrammatic form, the structure of *penicillin*. This expenditure was fully justified by the role of such a diagram in interpreting innumerable chemical reactions involving this compound – reactions that might, for example, have shown the way to an economical non-biological process of synthesis. That, surely, is what we mean by 'chemical knowledge'.

The *stereospecificity* of biologically active compounds is the major organizing principle of *molecular biology*. A typical protein molecule (Fig. 22) is 'coded' linearly as a succession of amino acid residues. This could be stored in a computer memory as a 'number' of, say, 2000 'bits'. But the catalytic power of such a molecule as an enzyme depends on the spatial configuration into which the chain naturally folds in the cellular medium. This, again, could be represented by another 2000 'bits' of data.[4] Yet such a method of storing the information would conceal the theoretical key to the activity of the enzyme – the relative positions of certain chemically significant sites on the folded chain. But these would at once be obvious to visual inspection of, say, a rotating stereographic map of the three-dimensional structure of the molecule.

Even in *physics*, material maps are often indispensable. For example, much of what is known about the electrical properties of a crystalline metal is summed up in a *Fermi surface* (Fig. 23), where each point represents a possible value of the *momentum* of the electrons in the metal. Although the *momentum space* in which this map is drawn is a mathematical abstraction, various numerical dimensions of the Fermi surface can be directly determined by experiment. Thus, the shape of a continuous surface of this kind for every metallic element fully represents the cognitive content of this field of science.

As for *mathematics* itself, it has long been recognized that a geometrical theorem is no more than a particular realization or representation of some underlying symbolic or verbal chain of argument. It

[4] Jacques Monod (1972) *Chance and Necessity* (London: Collins) p. 92.

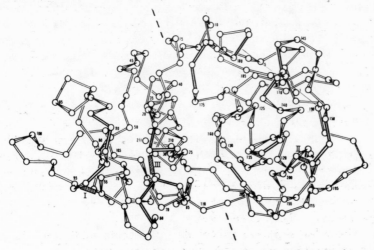

Fig. 22. The active site in papain is in the cleft defined by the broken line. From Drenth, J. *et al.* (1968). Structure of papain. *Nature*, **218**, 932.

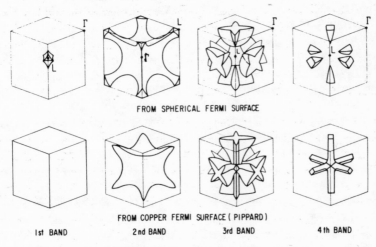

FROM SPHERICAL FERMI SURFACE

FROM COPPER FERMI SURFACE (PIPPARD)

| 1st BAND | 2nd BAND | 3rd BAND | 4th BAND |

Fig. 23. The folded Fermi surface for ordered Cu₃Au. Reproduced by kind permission of the General Electric Corporation.

is thus held, in general, that a geometrical demonstration is inferior to a more abstract 'analytical' proof. In reaction to Newton's Euclidean methods, the eighteenth-century French mathematicians even took pride in writing books on Analytical Mechanics without the use of diagrams. Yet we may suspect that, as the author and readers of such texts worked their way through the algebra, they did occasionally draw for themselves little sketches on scraps of paper.[5] The modern mathematician does not so deliberately set to work with one hand tied behind his back; but his task is complicated by the fact that much of modern mathematics is concerned with *topology*, which is precisely the general theory of abstract maps of all possible kinds. Theorems of great generality in spaces of many dimensions can only be illustrated very schematically by material maps in the Euclidean spirit.

4.2 *The map metaphor*

In emphasizing the similarity between a scientific theory and a map, something much deeper is being suggested than that scientific information is often stored and presented in diagrammatic form. Pursuing the metaphor further, we uncover many important characteristics of scientific knowledge in the cognitive realm,[6] where it is temporarily detached, so to speak, from observational factuality.

A map, for example, has to be drawn to fit the data in the surveyor's note books – information that is always incomplete and subject to error. In many of its details, therefore, the map may be no more reliable than a shrewd guess or a rough interpolation. In the same way, a scientific theory is an attempt to fit incomplete and imperfect experimental evidence (2.9, 3.3) and necessarily contains many uncertain or conjectural elements.[7]

But when we try to improve or correct a map – perhaps to accommodate new information – we note that it is *multiply-connected*: it cannot be significantly altered at one point without repercussions in the neighbourhood. The surveyor deliberately collects *redundant* data, so that the locations of the main features of the map are *over-determined*.

[5] For 'To pursue a geometrical strain of thought by pure logic without seeing constantly before my eyes the figure to which it refers is impossible at any rate for me'; Felix Klein quoted by E. Casirer (1956) in *The Problem of Knowledge* (New Haven: Yale University Press).

[6] I.e. in Popper's 'World 3' (1.4) or 'noetic domain' (5.5).

[7] For real maps there is a computer algorithm (D. G. Kendall 1975: *Philosophical Transactions* **279**, 547–82) for the 'Recovery of structure from fragmentary information', such as adjacency relations of the *Départements* of France. The metaphor does not stretch to assume that the same procedure may be applied to scientific theories.

Analogously, scientific knowledge eventually becomes a *web* or *network* (2.10) of laws, models, theoretical principles, formulae, hypotheses, interpretations, etc., which are so closely woven together that the whole assembly is much stronger than any single element.

This characteristic of any well-established body of scientific knowledge is often obscured by a conventional historical account of the initial phases of its discovery. Metaphorically speaking, the paths followed by the first explorers, passing by and recording many significant features of the landscape, are somewhat arbitrary, and may not cover the territory systematically. With only their notes to go by, we could imagine many alternative maps of the same region – just as some naive critics of relativity or quantum theory (6.6) suppose that their own homespun theories of space, time and matter need only be consistent with a few famous experiments, such as the Michelson–Morley experiment and the photo-electric effect, which happened to play an important part in the historical evolution of modern physics. This fallacy arises through ignorance of the innumerable detailed investigations[8] that have since criss-crossed the territory and definitely resolved any major ambiguity in the original map – in this case, the whole body of research in atomic, nuclear and sub-nuclear physics of the past 50 years.[9]

More information can be read from a map than was needed to construct it. In the same way, a well-ordered body of scientific knowledge is an endless source of reliable predictions (2.8) going far beyond the existing accumulation of observational data. Notice, moreover, that the information abstracted from a map may itself be 'theoretical', in the sense of generalized geographical relationships between the 'places' on the map, such as that Bath, Chippenham, Swindon, Didcot and Reading lie on the railway from Bristol to London. In a science such as physics, we may similarly deduce a wide variety of partial 'laws' or formulae that are implicit in the overall theory but which may never previously have been stated.

Many of these map-like characteristics of scientific knowledge could be said to arise from the fact that it is *structurally multi-dimensional*. To illustrate this point,[10] let us think of a one-dimensional map or *itinerary*,

---

[8] Kuhn's 'normal science' of course!

[9] For this reason, I do not personally favour an historical approach in teaching the first elements of an unfamiliar branch of science. A good grasp of the *true* history of his subject is an important intellectual asset for any scientist, but not at the expense of misconceptions as to its current validity. It is essential not to confuse romantic admiration – or envy – for the heroic achievements of past genuises with deference to – or rebellion from – their actual scientific theses.

[10] Which is made by Toulmin, *The Philosophy of Science.*

such as the list of railway stations between Bristol and London. Given the distances between successive stops, various further predictions could be made and confirmed concerning relative journey times. But this 'map' would be no guide to the consequences of an accident blocking the section from Bath to Chippenham; who would know of the alternative (if slower) route via Westbury? By analogy, a scientific theory representing only a *chain* of logical or empirical consequences – for example a time sequence of 'causes' and 'effects' – is weak for validation and barren for prediction. The valuable 'map' characteristics to which we have referred would scarcely be evident in such a theory.

This point is of the greatest importance, since it explains so much of the strange sense of unreality that scientists feel when they read books on the philosophy of science. It is abundantly obvious that the overall structure of scientific knowledge is of many, many dimensions – more, perhaps, than can be grasped by the human mind. The initial path to a new discovery (3.6) may be apparently 'one-dimensional' with no more reliable authority than a simple causal chain. But the strategy of research is to seek alternative routes, from other starting points, to the same spot (3.5) until the discovery has been incorporated unequivocally into the scientific map. The fact that this strategy seldom fails is unfortunate for purveyors of one-dimensional verbal analyses of scientific methodology, but is the source of the extraordinary reliability that scientific knowledge can often attain in practice.

The map analogy applies to other features of scientific knowledge considered as a body of 'theory'. Thus maps can be made on various *scales*, covering larger or smaller regions, with less or greater local detail. Scientific theories represent our knowledge at various levels of generality – broadly and abstractly, as with the laws of thermodynamics, or minutely and very specifically, as in the mode of replication of DNA. To the geographer, no confusion is occasioned by the fact that, say, the city of Bristol appears as a mere dot on the motorway map of Great Britain, and yet can spread and spread over many sheets of the 25-inch Ordinance Survey maps. To the scientist there is no confusion in treating an atom as a tiny sphere, in the kinetic theory of gases (2.2), yet knowing that its spectroscopic properties require that it should be thought of as a horribly complex cloud of interpenetrating electron orbitals buzzing round a central nucleus. The real question for the scientist (as for the geographer) is whether the level of generality is appropriate to the problem in hand and whether the details are then sufficiently accurate to solve it.

What we also recognize is that a *sketch* map can convey significant and reliable information, without being metrically accurate. What such a map represents of course is the *topology* of the relationships between recognizable geographical features – for example, the sequence of stations and their interconnections on the London Underground. In many fields of science, what we call *qualitative knowledge* has these characteristics – for example, the ethologist's account of the courtship behaviour of birds or baboons. Is such knowledge 'unscientific' because it is not quantitative – because the subway map does not, so to speak, show the actual positions of the stations by latitude and longitude. The question is, rather, whether the sketch or diagram correctly represents the significant relationship between identifiable entities within that field of knowledge – often a very moot point that cannot be resolved by mechanical counting or 'measuring'.

4.3 Pictures

The map metaphor for science preserves us from a vulgar fallacy – the tendency to conflate scientific knowledge with the material reality that it purports to describe. No sane person would suppose a map to be identical with the land that it represents.[11] In ways that we understand in practice (though one might be hard put to it to define formally) a map is necessarily an *abstract* representation, whose features are schematic, and quite unlike the objects from which they are derived except in, say, their mutual topological relationships. On the motorway map, the blob marked 'Bristol' resembles that city solely in that it lies on the line marked 'M5' between blobs standing for 'Gloucester' and 'Taunton'. When, scientifically, I write

$$Zn + H_2SO_4 \rightleftharpoons ZnSO_4 + H_2$$

I am not attempting to tell the whole truth about metallic zinc and sulphuric acid, in all its shimmering, fascinating detail (7.3).

But between the actual landscape, on the one hand, and its map, on the other, we may find an intermediate, information-loaded artefact – the *picture*. As we have seen (3.2), drawings and photographs play a very important part in scientific research. They are recorded by individual observers, transmitted from one to another, studied for evidence, and often stored in the scientific archives. Yet we reject intuitively the notion that a collection of such pictures, without theo-

---

[11] F. Capra (1975) in *The Tao of Physics* (London: Fontana/Collins) p. 94, points out that in Hindu philosophy '*Maya* is the illusion of taking the concepts for reality, of confusing the map with the territory.'

retical interpretation, would constitute a mature body of scientific knowledge.

Here is a subtle point: what is the difference between a picture and a map, or diagram? Give yourself the pleasure of standing before a great landscape painting. Or go to the technological limit of a coloured movie, projected on to an all-enveloping screen, complete with stereophonic sound and artificial aromas – perhaps, in fantasy the 'feelies' of *Brave New World*. It is clear that a picture is designed to give the semblance of reality, *as experienced by the individual observer*. At its best, it tells us about 'the world of our conscious experiences',[12] from which, eventually, scientific knowledge is to be derived.

This subjective or personal aspect is fundamental to picture-making as an *art*. The general faculty of pattern recognition enables most people to see much the same things even in a badly drawn picture (5.9). But coming to an artistic masterpiece, each person with a different life history and emotional expectation (7.10), there is no assumption that it must evoke the same thoughts or feelings in us all. Contemporary aesthetic theory insists that the artist should strive to convey his personal vision, but only to release in his viewers the richness of their own diverse responses. Indeed, the trend in the visual arts is continually away from the crude verisimillitude of a photographic likeness into the subtle depths of psychic evocation.

For the needs of science, however, these emotional characteristics, this evocation of a particular human situation, this individual viewpoint, cannot be completely reduced to a consensual communication. Even a carefully painted picture cannot be fully tested for its accuracy by a network of independent critical observers – too much depends on the spot where the artist stood, there are too many details hinted at by a squiggle or a blob of paint. In the very process of establishing agreement, we begin to simplify, we draw diagrams, we exchange arithmetical and logical communications and thus reduce the consensual elements in the picture to the more abstract, less subjective form of a map.[13] And when a taxonomist makes a drawing of a plant specimen (Fig. 4), he depicts it schematically, emphasizing character-

[12] I.e. the *mental domain* (5.5) or 'World 2' of Popper, *Objective Knowledge*.
[13] At an exhibition of the works of John Constable at the Tate Gallery (Spring 1976) a map of Dedham Vale was provided to show the points from which so many of the landscapes had been painted. This map incorporates all the significant geographical information provided by the paintings, establishes a consensual standard against which accidental or deliberate artistic discrepancies can be detected, and gives a further dimension of meaning to the pictures themselves as transitory, highly personal experiences of that particular landscape. But the map is a totally inadequate substitute for the paintings!

istic features such as the external outlines of the leaves, with little care for the realities of light and shade and perspective.

In thus distinguishing between a map and a picture, we must not be confused by the existence of mechanical processes of visual reproduction. A *photograph*, for example, may be an immensely potent item of scientific evidence (3.2). By eliminating the human error in the fixing of observational details, by its promptitude, ubiquity and multiplicity, photography is an indispensable scientific tool. But the camera is aimed, and the film is exposed, according to the intentions of the photographer – a conscious human being. If he is an artist, then he chooses to shoot precisely those scenes whose aesthetic or emotional intensity makes them 'pictures'; as a scientist, he automatically trains his lens to bring out the potentially consensual elements in what he observes. The ideal of scientific photography (e.g. the bubble-chamber tracks in Fig. 11) is to approach as closely as possible to a 'diagram' where the underlying conceptual structure is emphasized, stripped of all inessential detail. Indeed, the marvellous LANDSAT photographs of the surface of the earth[14] taken from a satellite and reproduced in 'false colours' to emphasize the differences between various terrains and crops, are almost perfect maps whose 'subjective' characteristics have been eliminated by putting the point of observation high in the sky, far above any human eye.

This whole question of the nature of scientific theories, and of the relationship between scientific knowledge and the cognitive contents of the individual human mind, is of extreme philosophical difficulty. The subject matter of science is so diverse, the conceptual powers of the mind are so uninhibited, that there is no way of fixing the elements of the relationship by generalized definition, in abstraction. Only analogical demonstration can bring it within our comprehension. But in the light of the map metaphor, we see that scientific knowledge is necessarily schematic and 'theoretical'. Because its contents must be consensual for a large and very critical community, it cannot represent all the miscellaneous, adventitious detail of actual life, as experienced by any single individual. This is what we mean when we emphasize the *objectivity* of science (5.6); it makes maps to inform us, not pictures to move us with pity and terror.

[14] See e.g. *Scientific American*, September 1976.

4.4 Paradigms

Science does not grow by simple accumulation. The carefully observed, criticized and theoretically schematized knowledge that is transmitted to the archive (6.5) is not thereafter hoarded in secret vaults; it becomes the free property of all men, including the scientists themselves, and is instrumental in the generation of further knowledge.

Nor is a scientific observer an inflexible machine, fully formed by his education. Being himself involved in the generation of new knowledge, he is continually revising his own creative and critical standards in the light of scientific progress. As the means become evident, as the possibilities present themselves, as new doors are opened by his own work or the work of other scientists, he constructs more sensitive apparatus (3.3), seeks to confirm recent predictions (2.9), applies new theoretical formalisms (2.4), reinterprets previous discoveries (3.6) or conceives new programmes of research. In other words, scientific activity is self-catalysing and self-correcting; it is governed by the outlook and directed towards the problems of its own day, as perceived by its human practitioners.

To illustrate this dynamical process, it would be necessary to penetrate into the obscure history of some particular branch of science, to show what information was potentially available to each research worker at the time, to note deficiencies of communication, and external stimuli of inspiration, to wonder at imaginative leaps and inexplicable blockages. The sources of invention turn out to be extraordinarily subtle and episodic (6.7), revealing little more than the diversity of human behaviour in unfamiliar circumstances.

Our immediate concern, however, is not so much with the psychology of *discovery* as with the sociology of *belief*. How does the scientific community react to the appearance in its midst of genuine new knowledge – in its ideal form, a well-ordered and convincing network of facts and interpretations, such as the theory of special relativity or Pasteur's clinching demonstration of the bacterial causation of disease.

After the initial period of scepticism and resistance, a major new scientific principle carries all before it. Having been the subject of intense research, having stood the test of many efforts at refutation, it acquires a highly reputable, almost unchallengeable status. It is the pride and joy of its creators, who are rewarded with recognition, who teach it with relish and who cannot resist imposing it inexorably on acquiescent juniors (6.2). To embed oneself mentally in the new theory, to demonstrate one's mastery of it, to make it the basis of one's

research,[15] is progressive and up to date. A whole new area of knowledge is quickly explored and mapped out as a consequence of the
'breakthrough'.

Here again, we need not go into the question whether the long-term
progress of science is ideally served by such waves of enthusiasm. What
we should note is that the new principle – a metaphorical map of some
corner of the world of nature – is rapidly *internalized* by every scientist
to whom it seems relevant (5.6). It is not just something that he reads
about in the scientific journals, or a technical device that he can pick
up, use and put down again as the occasion demands. As he solves
problems with it, teaches it to his students and argues with it amongst
his colleagues, he assimilates it as a *concept*, until it becomes a part of
himself. From the 1930s onwards, quantum mechanics, for all its
philosophical paradoxes, was not just a 'theory' that could be used,
if necessary, to explain atomic phenomena; to the atomic physicist,
quantum mechanics had become *reality* (5.10); it was no longer possible
to think physically in any other categorial language.

Thus, from a *scientific revolution*, evolves a new paradigm.[16] Or, in
the language of the visual metaphor (4.3), the *map* has become a
picture.

Arriving in an unfamiliar city, we consult a map. A route is calculated
by a succession of logical deductions. 'This place is called "Trafalgar
Square". Here is "Trafalgar Square" on the map. I propose to go to
"Buckingham Palace". On the map, "Buckingham Palace" lies West
of "Trafalgar Square", at the end of "The Mall". Therefore I must
look for a street called "The Mall" on the West side of this place', etc.
etc. With time, however, we learn our way around. The map needs
to be consulted less and less, until eventually we can work out the most
complicated journey, 'in our head' without reference to the printed
sheet. In the street scene, we observe not only buildings, vehicles and

[15] I.e. to participate in normal science, as defined by Kuhn, *The Structure of Scientific Revolutions*, p. 5: 'the activity in which most scientists inevitably spend almost all their time is predicated on the assumption that the scientific community knows what the world is like'.
[16] T. S. Kuhn, *The Structure of Scientific Revolutions*, p. 108. Much of the present chapter can be summed up in his words '[In its role] as a vehicle for scientific theory, the paradigm functions by telling the scientist about the entities that nature does and does not contain and about the ways in which those entities behave. That information provides a map whose details are elucidated by mature scientific research. And since nature is too complex and varied to be explored at random, that map is as essential as observation and experiment to science's continued development. Through the theories they embody, paradigms prove to be constitutive of the research activity. They are also, however, constitutive of science in other respects...paradigms provide scientists not only with a map, but also with some of the directions essential for map-making. In learning a paradigm, the scientist acquires theory, methods, and standards together, usually in an inextricable mixture.'

people, but also various clues as to our whereabouts, of which we are continually conscious. For us, now, any attempt to represent 'the semblance of reality, *as experienced by the individual observer*' (I quote 4.3, above) would have somehow to include this internalized map of our surroundings, which has become part of the 'picture' at whose centre we stand.

Once again, it is almost impossible to explain this point, except by analogy or example. If you ask the atomic physicist to tell you how he personally feels about quantum mechanics, you will only get back the bleak official map representation taught in the text-books. But as a result of long experience and the accumulation of many details, he has in truth 'learnt his way around' in that field, and has transformed the theory – wave functions, matrices, Dirac delta-functions and all – into a personal world picture.

Paradigm formation is fundamental to science as a social process. Without the personal commitment of individual scientists to the same world picture, there could be no communication, no basis for criticism, no criteria for consensuality. Mental imagery – the capacity to think out the consequences of this or that action without laborious calculation (5.4) – is an immeasurably powerful human skill, essential to scientific progress.[17] The aim of a professional scientific education (6.2) is to develop the ability to live at ease within the current scientific consensus – to 'think physically' (or 'chemically', or 'biologically') as we dons like to put it. Experience has shown that it is practically impossible to contribute to physics, or chemistry, or geology, or physiology until one has become a 'physicist', or a 'chemist', or a 'geologist', or a 'physiologist' by internalizing the paradigm of one's subject.

But there is a price to pay for this commitment. Each generation of scientists gives too much credence to its own paradigms. By his education, and by participation in 'normal science', the average research worker is heavily indoctrinated (6.2) and finds great difficulty in facing the possibility that his world picture might be wrong (4.5). Thus, for example, the idealized anatomist (3.2), who draws what he 'sees' in a dissection, is strongly prejudiced in practice by what he 'knows' he ought to see. The virtues of scepticism are extolled (5.6)

[17] 'We make for ourselves internal pictures or symbols of external objects and we make them of such a kind that the necessary consequences in thought of the pictures are always the pictures of the necessary consequences in nature of the objects pictured ...When on the basis of our accumulated previous experience we have succeeded in constructing pictures with the desired properties, we can quickly derive by means of them, as by means of models, the consequences which in the external world would only occur in the course of a long period of time, or as a result of our own intervention.' Heinrich Hertz *Principles of Mechanics* (1894: English edition 1899).

but are difficult to practise in the everyday activity of research. The expert – who ought to know better – forgets or mentally suppresses past revolutions of thought, and is often more credulous than the layman concerning the foundations of his knowledge.

A scientific paradigm seldom has the richness and diversity of a real-life picture. It derives from a theoretical or phenomenological map, which is necessarily a schematic representation of the knowledge that it categorizes. This is particularly true of the typical scientific curriculum (6.2), where the observational evidence is usually reported at second hand, where complicated theories have to be taught in simple form, and where difficulties and contradictions are swept out of sight in the name of academic tidiness. The student is likely to pick up a grossly over-simplified and uncritical view of his subject, which may take many years of harsh experience to unlearn. All too often, the paradigm in the mind of a practising scientist is almost a parody of the subtle, sophisticated, many-coloured picture that would be built up by thoughtful scrutiny of the scientific archive.[18]

These deficiencies need not deceive us when we try to depict natural phenomena about which we have much more than 'scientific' information. It is difficult (although, not impossible) to have completely illusory notions about tables and chairs, cats and dogs, cabbages and kings. We always have our own commonsense (5.10) and the use of our eyes and ears to put us right.

But there are realms of nature that are not accessible to unaided human perception, or that are very unfamiliar to everyday experience, where the only picture we can form comes to us through scientific research. The distant reaches of the Universe, the origins of life, the properties of sub-microscopic particles – perhaps, also, the farther shores of psychopathology (7.8) – can only be elucidated by extravagant instrumentation (3.4), by subtle logical inference (2.8), or by devoted investigation of the merest vestiges of evidences. Concerning such realms, imperfectly confirmed, highly speculative, or boldly genera-lized theories are easily formulated, and take hold of the imagination of scientist and layman alike.[19] Such theories may acquire widespread authority, not because they are well founded and reliable but because

---

[18] E. H. Gombrich (1960) in *Art and Illusion* (New York: Pantheon Books) referring to the art of caricature, comments (p. 336) on 'the willingness of the public to accept the grotesque and simplified because its lack of elaboration guarantees the absence of contradictory clues!'

[19] Gombrich quotes a Chinese treatise on painting (p. 269) 'Everyone is acquainted with dogs and horses, since they are seen daily. To reproduce their likeness is very difficult. On the other hand, since demons and spiritual beings have no definite form, and no one has ever seen them, they are easy to execute!'

they have no competition from other less consensual sources of knowledge or insight. Whether or not it is eventually validated by overwhelmingly convincing evidence[20] the 'scientific picture' presented by
this sort of theory is inevitably schematic and over-simplified. The
danger is that its limitations will not be adequately recognized, and
that it will be extrapolated recklessly into an all-embracing dogma.

As an example of such dogmatism consider the famous question of
the direction of the 'arrow of time'.[21] There is little doubt that the
microscopic equations of physics, both classical and quantal, can be
formulated in such a way as to make no distinction between the two
directions of time – into the future or into the past. For this reason,
it is argued that the actual direction of time in actual physical systems
is (somehow) arbitrary, or is imposed only by the form of the boundary
conditions. But not everything that can properly be said about time
is necessarily expressed by the way it appears as a variable in the
mathematical language of physics (2.7). There is no 'scientific' justification for ignoring our own subjective (and absolutely consensual)
experience that change in time is a one-way street.[22] The time variable
itself is a 'theoretical' construct, invaluable as a coordinate dimension
for our maps of natural phenomena, but not constrained in its real
properties by the logico-mathematical conventions of the differential
calculus. But of course there is a perverse romanticism in suggest-
ing that the 'scientific picture' of time contradicts the commonsense
experience of all humanity!

4.5 Fallibility

Any serious consideration of the credibility of science must face up
to the fact that even the 'hard' sciences are fallible. There is a tendency
amongst scientists to suppose that this is always occasioned by faulty
or incomplete information. The conventional model of scientific
change permits a false hypothesis to receive tentative support in the

---

[20] An example of 'induced realism' is to be found in Jerusalem. On the basis of
archaeological and historical evidence, a large model of the City of Herod was made
a few years ago in the grounds of a hotel. In this model, a hippodrome stands
between the Temple Mount and the City of David. More recent excavations have
clearly demonstrated that there was no hippodrome on this site: yet it is hard to
believe that this part of the beautifully detailed and realistic model is a figment of
the imagination.

[21] See e.g. K. G. Denbigh (1975), *An Inventive Universe* (London: Hutchinson) which cites
the relevant literature.

[22] 'For is it not the case that we all make the same judgements concerning the sequence
in which events happen?...This is very different from the subjectivity of various
sensations such as pain and pleasure'. Denbigh, *An Inventive Universe*.

early stages, when the evidence is fragmentary. But further research reveals defects and contradictions; after a period of confusion and conjecture, there is an intellectual *revolution*, in which the previous paradigm is replaced by a more satisfactory synthesis, which is substantially correct in the light of the knowledge available at the time.

Perhaps this is a fair account of the progress of science in many fields. Yet this representation of science as automatically and rapidly self-correcting is not justified by the historical facts. It must be emphasized that quite erroneous views can become firmly established as scientific knowledge and not be eliminated despite strong evidence to the contrary.

We all know, of course, about 'caloric' and 'phlogiston' and 'spontaneous generation' (3.3); but those, we say, were childish fallacies of a past era, and not characteristic of modern science, where the errors are local and short-lived. Consider, however, a scientific fallacy on the largest possible scale, which persisted for half a century, until about 1960, in the face of widely publicized counter-arguments and counter-evidence. It is no longer necessary, surely, to remind the reader that Alfred Wegener put forward the hypothesis of *continental drift* in 1912 on the basis of the excellent fit of the continental margins.[23] If this fit were not an extraordinary coincidence, then further evidence supporting the hypothesis was worth consideration. For example, the distribution of various animal and plant species in the southern hemisphere could be explained by radiative dispersion across Africa and South America as if the Atlantic Ocean did not exist. Wegener pointed, also, to geological evidence such as similar geological formations in North-East Brazil and West Africa that might once have been joined.

In many respects, Wegener's book is very convincing. Why, then, was his hypothesis rejected by the great majority of geologists for something like 50 years? How did the enormous fallacy of the fixity of the main continental land masses remain entrenched in geology for so long – accompanied, incidentally, by such fantasies as 'land bridges' that eventually sank without trace into the wide, deep waters of the South Atlantic?

The story is complicated: for example, an important part in reconstructing the paradigm was played by geophysicists who found quite new evidence for actual motion of the continents from the magnetism

[23] See e.g. *Debate about the Earth* by H. Takeuchi, S. Uyeda and H. Kanamori (1967: San Francisco: Freeman, Cooper & Co); *A Revolution in the Earth Sciences* by A. Hallam (1973: Oxford: Clarendon Press).

of the floor of the Atlantic Ocean (6.3). For our present purpose it is enough to draw attention to this striking example of a large scale fallacy persisting for half a century in a mature and sophisticated field of science. Throughout this period the basic facts and hypotheses were well known, and there was no serious attempt to suppress publication of 'uncomfortable' truths by either side (6.7). The scientific system was working quite normally; individual scientists conscientiously carried out their research, which was communicated according to the 'norms' of our model (1.3). Unfortunately, in this case the 'knowledge' produced by 'science' was not as reliable as we usually assume, and the 'map' of Earth History in the archive was hopelessly misleading.

It is instructive, also, to note that the main arguments against continental drift came from mathematical physicists, who could easily show that the tidal mechanisms that Wegener had suggested were quite insufficient to bring about such large effects. But they failed to realize that this did not rule out some *other* mechanism which nobody had yet thought of. The epistemological irony is that the geologists – supreme experts in the observation and interpretation of visual patterns (3.2, 4.1) – rejected first-rate evidence from fossils, rocks and landscapes because they thought it was in conflict with quantitative mathematical reasoning (2.4) which few of them really understood. This episode shows that pattern recognition is every bit as reliable as a source of consensible knowledge, and as a means of arriving at a scientific consensus, as discourse in the logico-mathematical mode.

5
The stuff of reality

'When one reflects upon what consciousness really is, one is
profoundly impressed by the extreme wonder of the fact that an
event that takes place outside in the cosmos simultaneously
produces an internal image, that it takes place, so to speak, inside
as well, which is to say: becomes conscious.'

C. G. Jung

5.1   *Perception*

Scientific knowledge is made by and for people. Its contents and
quality depend on the powers and functions of scientists as *observers*
(3.1), as *communicators* (2.1), as *assessors* and *assimilators* of knowledge
(4.4), and eventually as *believers* (5.5) and as *authoritative experts* (6.5).
To grasp the meaning of science, to judge its credibility, we must look
more deeply into the human mind and investigate its capabilities and
limitations. This is the theme of the present chapter.

Of these capabilities, the most primitive is the faculty of *perception*.
The epistemological issue cannot be faced without serious attention
to the means by which the natural world is 'observed' and thus
brought into *cognition* (3.2). In other words, we must consider the most
elementary capacity of a human being – to respond intelligently to
some feature of the environment.

As before (3.1), we talk about *visual* perception, taking for granted
the analogous senses of sound, touch, smell and taste. There is no call
to be fastidious about nomenclature; our model of science is really
much too general and coarse-grained for nice distinctions between
'perception', 'pattern recognition', 'seeing', 'observing', etc. The
important point is to realize the relevance of the psychology of perception to the philosophy of science and to the overall credibility of
scientific knowledge.

This quite familiar and ordinary human skill is shared to a
considerable degree by all the higher animals. Apes, elephants, and
dolphins have different sensory facilities and interests from men, but
can make comparable discriminations, even though they cannot
describe the features of a pattern, or name the objects therein
perceived. All that they lack is the specifically human ability to
transform the received information into a message for social
communication.

95

Fig. 24. From Harmon, L. D. *Scientific American*, November 1973.

Instrumentally, 'the human viewer is a fantastically competent information processor'.[1] The number of 'bits' of 'information' in Fig. 24 is tiny; yet when the sharp edges of the squares have been filtered out, the blobs of shading coalesce into an identifiable face. A very imperfect representation can thus be *recognized* – i.e. related mentally to previous incidents of patterns which in some ways it significantly resembles.

What is the underlying mechanism of perception? This is still one of the major mysteries of science. Research has made considerable progress in some directions, but there is no clear understanding of the process as a whole. From *neurophysiology* we learn that photochemical receptors ('rods' and 'comes') on the retina respond to light, and send nerve impulses through the optic nerve to the brain. But even at this stage we find mini-processors, in the layers of nerve cells behind the retina, that detect specific features of the visual pattern, such as 'edges' between light and dark regions, or that respond only to changes of illumination.

The physiological machinery of eye and brain can thus 'recognize' some of the characteristic elements of the supposedly abstract knowledge summed up in the languages of geometry and dynamics. The lines that make up a geometrical diagram may not, after all, be highly contrived and artificial intellectual constructions, but can be grasped as visual percepts and transmitted as elements of information to the brain. Mathematicians and their attendant historians treat 'velocity' as a theoretical concept, requiring the separate abstractions of 'space' and 'time', which are then divided one by another; yet there is evidence from physiology that 'motion' is directly perceived by the human instrument without the intervention of an elaborate internal

[1] L. D. Harmon, *Scientific American*, November 1973.

calculation. These elementary facts of biology carry us very far from
the philosophical analysis of perception as primarily the intellectual
processing of separate signals about 'patches of colour' projected upon
the separate cells of the retina.

The fate, within the brain, of the signals leaving the optic nerve is
little understood. Neurosurgery and other clinical evidence suggests
that various regions of the brain carry out specific functions, whose
role is essential in the overall process of perception. It is known, for
example,[2] that lesions of the 'secondary visual zone' can lead to a
peculiar pathological symptom known as 'visual agnosia'; the patient
is unable to recognize familiar objects or patterns in a simple picture
that has been 'crossed out' by a few lines drawn across it. Again the
'tertiary zones of the parieto-temporo-occipital cortex' are concerned
with the integration or synthesis of perceptions; damage in this region
can lead to the inability to tell the time by observing the hands of a
clock whose face is not numbered, or to find ones bearings on a map.
'Frontal lobe' lesions give rise to failure to 'understand' a thematic
picture: faced with a sketch of, say, a boy falling through broken ice
on a pond, the patient throws out wild conjectures as to 'what is going
on'. An interesting observation is that 'Kanji' – the Japanese phonetic
symbols – can be read by a patient whose brain damage prevents the
reading of the Chinese-style ideograms of 'Kana'.[3] These facilities for
pattern recognition thus employ different mechanisms or nervous
pathways in the brain.

Without going deeper into this highly specialized and, as yet, fragmentary body of scientific knowledge, it is clear that one cannot really
separate 'perception' from the more general functions of intelligence.
The visual information reaching the brain from the eyes is inevitably
transformed into other signals. These transformations are not themselves invariant. The apparent locales of various higher mental processes move about in the brain with physical growth and the acquisition
of experience; thus, for example, as the action of writing is learnt, it
becomes a 'single kinetic melody' with its own characteristic region.[4]

[2] See A. R. Luria (1973) in *The Working Brain* (London: Penguin Books).
[3] A. Yamadori (1975) *Brain*, **98**, 231.
[4] Considering the 'system' as a whole, Luria remarks further on the relationship
between mental action and the external social aids of language and symbol. 'Whereas
higher forms of conscious activity are always based on certain external mechanisms
(good examples are the knot which we tie in our handkerchief so as to remember
something essential, a combination of letters which we write so as not to forget an
idea, or a multiplication table which we use for arithmetical operations) – it becomes
perfectly clear that these external aids or historically formed devices are *essential
elements in the establishment of functional connections between individual parts of the brain*,
and that, by their aid, areas of the brain which previously were independent become
the *components of a single functional system*'. Luria, *The Working Brain*, p. 30.

We need not rely heavily on the *particular* facts about the brain. The main lesson of neurophysiology and of cognitive psychology is that pattern recognition is deeply embedded in the complex structure and functioning of the brain and mind. It is also highly dynamic and historical, dependent fundamentally on memory, and closely connected with the capacity for 'problem solving' that is the hallmark of intelligence in animal or man.[5] This is a point to which we shall return when we consider the psychological development of the individual from childhood (5.8).

5.2 *'Artificial intelligence'*

Our understanding of any natural phenomenon does not always come from the step-by-step accumulation of experimental data and observation, clarified by successive stages of interpretation. Much of science has arisen through theoretical conjectures, model building and confirmation of predictions (2.8). This strategy has been applied enthusiastically to the problem of visual perception. Since elementary conceptualized theories could prove almost nothing, the characteristic approach has been to try to *simulate* human perception by an opto-mechanical machine linked to an electronic computer (7.7).

It would be wrong to suggest that research into *artificial intelligence*[6] has made no progress. Computer programs have been constructed that will, for example, do a very reliable job of 'recognizing' the various items in a finite set of conventional patterns such as well-drawn letters of the alphabet in random sizes and orientations. Many aspects of the processes of reducing visual patterns to coded messages conveying the (possibly) significant features are now well understood. But the wisest assessments of the actual achievements of this research[7] are not optimistic about progress to the desired goal of simulating human

[5] The 'hypothesis matching' model of vision is expounded with characteristic grace by E. H. Gombrich in *Art and Illusion*. R. L. Gregory (1971) *The Listener*, **86**, 359, sums it up pithily. 'Perception cannot be described purely in terms of stimuli eliciting responses, for we respond to features of objects not being represented to the eye or other senses. We expect an ice cream to be cold, though the (stimulus) image of the ice cream in our eyes is not cold. We act towards a table as though it had four legs, though two are hidden. In short, behaviour is appropriate to what is not available to the senses. Further action is aimed, not at the present but at future events... Perception can, I think, be described by saying that from sensory data we build up hypotheses of the external world of objects.'

[6] As expounded, for example, by E. B. Hunt (1975) *Artificial Intelligence* (New York: Academic Press).

[7] E.g. J. Weizenbaum (1976) *Computer Power and Human Reason* (San Francisco: Freeman).

perception by artificial means. It seems, rather, that the genuine difficulties of the whole problem are becoming more evident and look less surmountable now than they did 10 or 20 years ago.

The superiority of the human eye and brain over any machine is vividly exemplified by the technique of 'interactive' computation, where a computer is programmed to present on a screen various 'pictures', of a product under design. These pictures are derived from an arithmetical listing of the design parameters of the product, which can be modified after visual inspection by the engineer, architect, car designer, or other technical expert. In the same way, the 'automatic' scanning of bubble-chamber photographs (3.2) cannot entirely dispense with a human operator to recognize or discriminate between the various types of event that might happen to be recorded in the photographs. This is not to say that a system of photocells coupled to quite a modest microcomputer cannot far exceed human powers, in speed and accuracy, in very stereotyped tasks such as sorting clean from dirty bottles. But the fact is that we do not have anything approaching a computer program or other mechanical device that can match the performance of a human being in quite ordinary processes of pattern recognition and discrimination. And from the study of the neurophysiology and psychology of perception (5.1) there is little to encourage the belief that this immensely complicated problem will shortly be solved by some technological *tour de force*.

5.3 Extra-logicality

Science depends fundamentally on human powers of perception, recognition, discrimination and interpretation. The scientist as observer or communicator is an indispensable element of the knowledge system. But these powers have not been simulated by an artificial, non-human device: there is no computer program, no formal algorithm, no string of logical operations, to which these processes are equivalent, or to which they can, in any practical sense, be reduced. *Therefore* – and this is one of the most important characteristics of the 'consensibility' model of science (1.3) – *scientific knowledge cannot be justified or validated by logic alone.*

This proposition – anti-positivist, anti-inductionist – is now well entrenched amongst philosophers of science (1.3) even if it is not yet universally accepted. Yet it seems to conflict with the striving for rigour, exactness, predictive certainty that is fundamental to the whole research enterprise. If science is extra-logical, then what is all the fuss about? We might as well record our memories of opium dreams as the output dials of automatic spectrophotometers.

The paradox is resolved by reference back to Chapter 2. Logicality (ideally two-valued; at worst three-valued: 2.6) is a necessary property of the *communications* that carry, and embody, scientific knowledge. An ambiguous or self-contradictory message (2.1) cannot be the basis for a consensus; the effort to eliminate illogicalities from the scientific archive can never be relaxed; a large proportion of scientific research is devoted almost exclusively to this end.

But the formal requirements that make a message unambiguous do not guarantee its *truth*. This is a trivial principle of logic, on which all the problems of epistemology seem to hinge. What we can now see is that the question of the credibility of science cannot be decided by reference simply to the contents of the archive (4.2), or by perusal of the innumerable communications that pass between scientists (4.2). According to temperament, one may be impressed by the coherence of well-established theories, or horrified by the contradictions of knowledge in the making. But the doubtful links in the chains of scientific certainty lie elsewhere, in the minds and thoughts and brains, perceptions and cognitions, of the many human beings who cooperate to create our maps of the material world.[8]

The assertion that human perception is extra-logical is not, of course, amenable to absolute proof. There is nothing in principle against the conception of a robot that could replace a human being as a 'scientist' in our system.[9] But this is still a distant goal, an ideal, whose potentiality derives from grossly over simplified maps (4.4) of brain function rather than from reliable knowledge of present day capability. To invoke this hypothetical device, which would presumably ensure the 'objectivity' of science by filtering and censoring our intersubjective messages and personal observations, is to give ourselves over to a metaphysical principle that is scarcely distinguishable from the 'God' of the limericks 'who was always about in the Quad' to

[8] We *must* reject the proposition that scientific epistemology 'need not concern itself with the very delicate questions involved in the discussion of the philosophy of perception and "our knowledge of the external world".' R. B. Braithwaite (1953) *Scientific Explanation* (Cambridge University Press).

[9] The criterion for success in this enterprise would be, as in the famous Turing criterion for an intelligent machine, that it would seem to be indistinguishable from a human being in a variety of ways. As things stand at present, 'a computer can only be said to be believing, remembering pursuing goals, etc., relative to the particular interpretation put in its motions by people, who thus impose the Intentionality of their own way of life on the computer' – D. C. Dennett (1969) *Content and Consciousness* (London: Routledge & Kegan Paul) p. 40.

observe, and ensure the continued existence of, the 'stately old tree'.[10] The question of the credibility of science must be answered in the light of our present intellectual and material resources, not according to those conjectured for a far-distant future.

5.4 *Intuition*

Even the scientific observer (3.2, 5.1) is not a camera (4.3). In due course, he must communicate his experience to other scientists; but between the pattern of excitations on his retina and the uttered message there occur many stages of internal neural manipulation. To these *mental processes* we have no direct access, except by introspection and highly circumstantial experimental inference. Yet they play an essential role in science, not simply by the extra-logicality that they introduce in the most elementary acts of perception (5.3) but as the primary elements of individual human cognition. The concept of 'knowledge' is entirely meaningless in the absence of some mental acts in the mind of some knower.

The ubiquity of intuition[11] in science can scarcely be overlooked, yet its philosophical status and epistemological consequences are notoriously controversial. How can we minimize the doubts this could evoke in the overall reliability of science without returning to the unrealistic positivist model which we have taken such pains to reject?

The only personal access we have to such processes is introspective, looking back inside the 'black box' of the mind from the message-uttering output. A familiar style of intuitive thinking is that of the mathematician who, as he puts it, is suddenly 'struck' by the solution

---

[10] (Anonymous):

> There was a young man who said, 'God
> Must find it exceedingly odd
> That that stately old tree
> Should continue to be
> When there's no one about in the Quad.'

To which comes the reply (attributed to Ronald Knox):

> Dear Sir; your astonishment's odd.
> *I* am always about in the Quad,
> So that stately old tree
> Will continue to be,
> Since observed by, yours faithfully, God

[11] Encompassing a diversity of mental processes, such as those summarized by M. Bunge (1962) in *Intuition and Science* (Englewood Cliffs, New Jersey: Prentice-Hall) p. 90, 'In the language by means of which we speak of science "intuition" designates modes of *perception* (quick identification, clear understanding and interpretation ability), *imagination* (representation ability, skill in forming metaphors, and creative imagination) *inferring* (catalytic inference) *synthesizing* (global vision), *understanding* (common sense), and *evaluating* (phronesis).

of a problem on which he has long been working.[12] How is it that he can arrive at the conclusion of what would seem to be a long sequence of mathematical steps without apparent conscious effort? The answer is not to be found in some extra-sensory power of the brain, but in his long experience of the outcome of the various symbolic operations that are his daily intellectual companions.[13] He is thus able to jump over many tedious intermediate steps, and arrive, as if by magic, at an astonishing outcome. Such enormous intellectual leaps are as baffling to the layman as the quite unsuspected mathematical theorems (2.4) on which they are based, or to which they give rise. But like the outcome of elementary mental arithmetic, or of a skilfully played hand of Bridge, they are not extra-logical, and are eventually validated by explicit mathematical argument, in the public language of external communication. We are not here concerned with the mysterious factor of *imagination* by which the creative mathematician, scientist, or other thinker brings to mind precisely the sequence of steps that will solve his problem: all that we need to realize is that this sort of thinking merely internalizes the characteristic properties and relations of concepts and operations that are already known and justifiable in public. The discoveries of pure mathematics and theoretical physics are 'mapped' coherently in the archives of science (4.2), and thus become living 'pictures' (4.4) in the minds of those who have learnt them.

The problem of the epistemological status of mental operations in a highly consensible language such as mathematics (2.3) is not really different from the more general issue of the meaning and validity of statements in any natural language. We shall try to keep out of the great forests of philosophical discussion that have sprung up around this ancient question, although it cannot be entirely ignored (5.8). If language could not be understood, if it had no meaning, then the concept of intersubjective communication would itself be null!

What is more to the point is the convincing evidence from psycholinguistics[14] that verbal messages are transformed, within the mind, into logically equivalent forms without conscious analysis. The sentence 'She bought a pig from him' is comprehended effortlessly as conveying the same message as 'A pig was sold by him to her'. The

---

[12] Cf. section 10 of Chapter 5 of Polanyi, *Personal Knowledge*.

[13] 'Seeing an object x is to see that it may behave in the ways we know x's do behave': N. R. Hanson, *Patterns of Discovery*, p. 22. 'The specialist in abstract structures, such as groups, evolves an intuition in handling them, which is a learned way of saying that he becomes *familiar* with such abstract structures': Bunge, *Intuition and Science*, p. 70.

[14] See e.g. *Language* edited by R. C. Oldfield and J. C. Marshall (1968: London: Penguin Books) – especially the chapter by D. McNeill.

limitations of this 'central processing unit' are easily tested by logically and grammatically sound sentences which are too complex to be processed 'intuitively' without mechanical aids (2.4). It seems obvious that the unconscious symbolic, operational, or 'conceptual' transformations of the mathematician are quite analogous to those undertaken in the mind of the merest oaf who can speak intelligibly in his native tongue, and are certainly not more mysterious. A capacity to 'understand' – i.e. act appropriately towards the contents of – a verbal communication seems as truly human as the powers of pattern recognition that we have already discussed at length (5.1). Indeed, these powers may be closely connected (5.8).[15] It would be ironical if the *Venn diagram* – a 'pictorial aid' to the representation of logical relationships – were in fact shown to be isomorphous with the brain patterns by which such relationships are actually comprehended.

But 'scientific intuition' includes many processes that cannot be intelligibly communicated from one scientist to another, except by hints, 'hand waving' or other extra-logical means. Polanyi refers to 'ineffable knowledge', such as the experienced surgeon's grasp of the topographic anatomy of the human body.[16] The numerous sectional representations of this topography would be of formidable complexity, yet the surgeon 'knows his way about in it' with inarticulable intellectual skill. The familiar processes by which we build 'pictures' of the world from subjective experience or from objective 'maps' (4.4) are of this mysterious character. We have the ability, it seems, to *visualize* a pattern that is not immediately before our eyes, and to subject it mentally to various transformations whose outcome may be explored and noted.

There is little doubt of the great power of *iconic thinking*[17] in the practice of science,[18] especially at the stage where theories are originally conceptualized. But this stage is not immediately relevant to the epistemological problem (but see: 6.7), so that this strongly

[15] There is evidence for a close neurophysiological connection between the mental machinery involved in dealing with spatial orientation, with arithmetic, and with logico-grammatical structures (Luria, *The Working Brain*).

[16] Polanyi, *Personal Knowledge*.

[17] H. R. Harré (1975), in *Determinants and Controls of Scientific Development* edited by K. D. Knorr, H. Strasser and H. G. Zilian (Dordrecht: Reidel) p. 259, defines an 'Icon' as a 'non-verbal image (often a pictorial or quasi-pictorial symbol) which is a vehicle for thought or a bearer of a concept, but, unlike words and such constructions, is projectively related to reality'.

[18] 'For me it is not dubious that our thinking goes on for the most part without use of signs (words), and beyond that to a considerable degree unconsciously'. A. Einstein, quoted, along with much other interesting material on the same theme, by G. Holton (1974) in *The Interaction between Science and Philosophy* edited by Y. Elkana (Atlantic Highlands, New Jersey: Humanities Press).

intuitive factor is not necessarily a serious threat to the credibility of
well-established scientific knowledge. Is it, for example, quite different
in kind from the more mundane 'logical intuition' to which we have
referred above?[19] Is it not, perhaps, simply another variant of the
'cognitive faculty', that encompasses all forms of pattern recognition,
from visual perception to the comprehension of language?

Our touchstone of reliability must be, once more, consensibility
leading to consensus. The intersubjectivity of verbal intuition is easily
tested by the exchange of messages. The difficulty with iconic thinking
is precisely that it is inarticulable, and cannot be communicated by
direct representation. Diagrams, sketches and other visual messages
play an important part in science (3.2, 4.1) but they are often inadequate to convey subtleties and complexities that may be apparent to
'the mind's eye'. A pattern which is thus imagined is not a permanent
object in the material world, which may be presented for recognition
to several observers in turn. Must it be forever relegated to the
extra-scientific sphere, interesting, useful, but not worth of belief.[20]

We may be comforted a little by the observation that this form of
intuition is indeed a universal faculty of the human mind. It is
fundamental to learning processes in childhood[21] and to the problem-
solving powers of all intelligent beings. There is evidence that the
deaf-mute, without the resources of inner dialogue in a natural language, can perform sophisticated mental operations of this kind,[22]
whilst the blind child can reproduce a very good drawn image map
of the house in which he lives.[23] It seems scarcely possible that a

[19] As suggested, for example, by C. G. Jung (1956) in *Symbols of Transformation* (London:
Routledge & Kegan Paul) p. 18, 'We have, therefore, two kinds of thinking: directed
thinking, and dreaming or fantasy thinking. The former operates with speech
elements for the purpose of communication, and is difficult and exhausting; the latter
is effortless, working as it were spontaneously, with the contents ready to hand, and
guided by unconscious motives. The one produces innovations and adaptation, copies
reality, and tries to act upon it; the other turns away from reality, sets free subjective
tendencies, and, as regards adaptation, is unproductive'. But the clinical goals of
psycho-analysis are not those of natural philosophy.
[20] In the words of M. Bunge (*Intuition and Science*, p. 111) 'the various forms of intuition
resemble the other forms of knowing and reasoning in that they must be *controlled*
if they are to be useful. Placed between sensible intuition, perception and pure reason,
intellectual intuition is fertile. But out of control it leads to sterility' and, p. 113:
'*Intuition is fertile to the extent that it is refined and worked out by reason*...Fruitful
intuitions are those which are incorporated in a body of rational knowledge and
thereby cease being intuitions.'
[21] See e.g. J. S. Bruner (1966) in *Learning about Learning* (Washington DC: US Dept. of
Health, Education and Welfare).
[22] J. Piaget & B. Inhelder (1969) in *Experimental Psychology; its scope and method:* VII.
Intelligence, edited by P. Fraisse & J. Piaget (London: Routledge & Kegan Paul) p. 200.
[23] *Image and Environment*, edited by R. M. Downs and D. Stea (1973: Chicago: Aldine)
p. 56.

similar mechanism is not at work in the brain of an intelligent animal, such as an ape, for example, when it stands on a box to knock down a banana with a stick, or a dolphin, when it skillfully tosses a ball through a hoop with a nudge of its head.

But the contents of the mind of the pre-linguistic child or animal can come only from its own individual experience. The intuition of the scientist has a much wider range of action, since it can operate on every item of information, every map or picture, every proposition or formula, to which it has access through languages and other media of communication. The flow of messages is not one way, from phenomena in the material world, through perception in the observer, into the social archive. The body of knowledge thus attained by the group is introjected back into the mind of each scientist, not simply as a fixed paradigm (4.4), but as further material for cognitive transformation, conscious ratiocination, or unconscious, intuitive reassociation and reformulation (6.7). The obstacles to consensibility and consensuality appear to lie less in the deficiencies of our personal intellectual powers than in the characteristics of the social groups to which we belong, and from which we draw our sustenance as thinking beings.

5.5 *Action and belief*

For all its subtleties and marvels, our scientific system does not necessarily tell us the *truth*. The social model (1.3), with its experimental instruments (3.1–3.4), its perceptive and mentally dynamic scientists (5.1–5.4), its media of communication (2.1), its theories, maps and pictures (4.1–4.4) might be no more than an entertaining toy, the apparatus of a very large and elaborate game,[24] with no function beyond the fun of being in it or watching it being played.

To connect with the fundamental question of this book, we must set science in the larger context of human *action*. The 'scientists' of our model belong to and are answerable to society in general, to which scientific knowledge itself ultimately belongs. Scientists are merely specialists in the acquisition of knowledge, which is only one of many human pursuits – living and loving, eating, drinking and making merry, buying and selling, making and mending. To ask whether science is to be *believed* is to ask whether there is valid connection between scientific knowledge and action in other spheres.

By what mechanism could there be such a connection? In elabor-

[24] E.g. *The Glass Bead Game* of Hermann Hesse (1970: London: Jonathan Cape).

ating our social model of science, we have identified three character-istic *domains*.[25] As organisms we exist in the *material domain*, and are acted upon by an 'external world' which we claim as scientists to observe and interpret. Within each one of us there lies a segment of the *mental domain* whose operations are perceptions, cogitations and other 'internal' processes of thought. By means of language and other media of communication we are in touch with the *noetic domain*[26] which is collectively created and maintained as a social institution. The material domain is itself a dimension of human action, and there is an ineffable machinery of conscious and unconscious processes in the mental domain governing our every move. But when we ask about *belief* in *science*, we are asking how this particular component of the noetic domain is related to other human pursuits. Such an influence could only flow through the *intellectual interface*, i.e. from the public archive of scientific knowledge by one or other of the communication media of science, into the mental domain of the individual human actor. In other words, a *belief is the internal mental counterpart of a message received from the noetic domain and capable of influencing personal action.*[27]

But the scientific message is not prescriptive. It is not an order from

[25] These are, of course, identical with World 1, World 2 and World 3 of Popper (*Objective Knowledge*). But since he is by no means the first philosopher to delineate these categories, there is no obligation to follow his clumsy and unevocative nomenclature.

[26] This slightly archaic name seems appropriate for what Polanyi (*Personal Knowledge*, p. 375) so eloquently defines when he writes: 'I shall regard the entire culture of a modern highly articulated community as a form of superior knowledge. *This superior knowledge will be taken to include*..., beside the systems of science and other factual truths, *all that is coherently believed to be right and excellent by men within their culture*...Only a small fragment of his own culture is directly visible to any of its adherents. Large parts of it are altogether buried in books, paintings, musical scores, etc., which remain mostly unread, unseen, unperformed. The messages of these records live, even in the minds best informed about them, only in their awareness of having access to them and of being able to evoke their voice and understand them. And this leads us back to the fact, implied in describing science as superior knowledge, that all these immense systematic accumulations of articulate forms consist of the records of human affirmations. They are the utterances of prophets, poets, legislators, scientists and other masters...'

Within this domain, Polanyi characterizes science more specifically as a '*coherent system of superior knowledge, upheld by people mutually recognizing each other as scientists, and acknowledged by modern society as its guide*', which is not very different in practice from World 3 of K. R. Popper (*Objective Knowledge*, p. 73) 'the logical contents of books, libraries, computer memories, etc'. The essential feature of this 'world' that it consists of ideas that have been *communicated* and therefore lie in the *public* or social realm is made by I. C. Jarvie (1972) *Concepts and Society* (London: Routledge & Kegan Paul) p. 152.

[27] Some of our beliefs arise directly, we say, 'from personal experience'. But this impression of individual autonomy neglects the internalization of noetic elements in consciousness itself; we accustom ourselves to rehearsing messages and public arguments within the privacy of our own thoughts, thus temporarily short circuiting the social processes of the systems.

a social superior, nor a moral imperative. The influence of knowledge over action arises from its power of prediction (2.8). To take a decision is to choose between branching paths into the future, in the light of their imagined or calculated end points. With the aid of the world map of science (6.5), we extrapolate with modest confidence into the near future, and hope to arrive at our desired goal. Scientific knowledge is reliable, science is to be believed, to the extent that this confidence is justified – not, of course, simply in the personal experience of each one of us, but as a consequence of the way in which it has been created by society for that purpose, and provides us with wise and prudent counsel.

It is, of course, perfectly possible to believe in science without any intention or opportunity to make that belief a basis for decision. The map of the universe presented to us by astronomy and cosmology is practically useless as a means of predicting the outcome of any action on the human scale of space and time. Within the limitations of the local laws of physics, and the state of things here on earth, the distant cosmos could be quite different without making a hap'orth of difference to the work, wealth and happiness of mankind. But the fact that scientific research spreads far beyond the immediately applicable, and that the generation of scientific knowledge becomes an end in itself, does not imply that the concept of belief can be disconnected from action. And the rhetorical force of a successful and subtle prediction in validating a scientific theory is intimately related to the practical need we have for reliable guidance into the darkness of an unknown future.

5.6   *Objectivity and doubt*

The primary foundation for belief in science is the widespread impression that it is *objective*. In contradistinction to what might be called 'subjective' it is '*knowledge without a knower: it is knowledge without a knowing subject*'.[28] The naive notion of the validation of science by a non-human robot censor has already been dismissed (5.3). The objectivity of scientific knowledge resides in its being a social construct, not owing its origin to any particular individual but created cooperatively and communally.[29] For all the glorification of individuals for

---

[28] Popper, *Objective Knowledge*, p. 109.
[29] This is expressed very clearly by D. Bloor (1976) in *Knowledge and Social Imagery* (London: Routledge & Kegan Paul) e.g. p. 86, 'things which have the status of social institutions are perhaps intimately connected with objectivity...perhaps that very special third status between the physical and psychological belongs, and only belongs, to what is social...the theoretical component of knowledge is precisely the social component'.

their great discoveries, the final product, the established theory and its confirmatory evidence, belongs, as we say, 'to humanity'. Newtonian physics establishes Newton as a great scientist; the mere fact that Newton discovered it does not make Newtonian physics a great science.

Scientific knowledge shares these characteristics of 'extra-subjectivity' and anonymity with other segments of the noetic domain – artistic convention, political prejudice, social tradition, etc. Although these are elements of 'superior knowledge' in which most people unreflectedly believe, we do not usually regard them as 'objective'. This epithet is applied to well-established scientific knowledge because of its great coherence and maximal consensuality (7.10). We know that it has been through a thorough process of criticism by well-motivated and skilful experts who have not been able to detect any significant errors. We suppose that we could ourselves test it for internal logical contradictions or for external observational consistency. The very fact that it now belongs to the consensus of the scientific community implies that it is 'believed' by its most expert potential critics (6.3); who would we be not to conform with their faith? In such circumstances, there is little space left for independent intellectual manoeuvres!

The objectivity of well-established science is thus comparable to that of a well-made map (4.2), drawn by a great company of surveyors who have worked over the same ground along many different routes. In the first instance, this map may not seem to agree with the little bit of the world that we see for ourselves; but with the experience of travel, in the absence of unsettling counter-evidence, we come to accept its characteristic features. Eventually it gives its own semblance to our picture of the world (4.4). The consonance thus achieved between our mental representation of our surroundings and our every movement in them is of the very essence of the relationship between well-founded belief and action. This is the basis of our confidence that we live in an 'objective' environment whose existence is independent of our own perceptions and conceptions.

Such an attitude towards science is widespread in our society. And yet the demons of doubt cannot be driven away. Science is demonstrably fallible. Firmly held and widely taught scientific principles have been found to be completely erroneous (4.5): '40 million Frenchmen' *can* be wrong. Just as the author of a literary fantasy may supply us with an imaginary map which evokes all the normal mental responses to an objective landscape, so a whole body of science may be an illusion masking an altogether different reality. There is nothing in the cognitive apparatus of the human mind, no one in the community of scientists, that can protect us from error or uncertainty. The best we can do, it seems, is to be eternally critical, eternally vigilant, eternally sceptical.[30]

An attitude of total *scepticism* is not philosophically untenable;[31] indeed, this attitude is easier to support argumentatively than the naive positivist extreme of accepting science as the naked truth. But taken absolutely, it is a negative, sterile attitude, paralysing the mind and will, and antipathetic to science itself.[32] Either our whole discussion stops at this discouraging conclusion, or we armour ourselves with a general commitment to 'the scientific attitude' in the domains of knowledge and belief (6.4) and sally out against more delicate questions than the crude totality of 'Yes' or 'No' to 'Is Science to be Believed?'. We should ask to what extent we should believe in science (6.5); we should enquire into its practical limitations (6.7); we should inform ourselves carefully as to the credentials of competing systems of belief (5.10); we should explore cases where a scientific consensus is still in the making, or where unsettling doubts have been expressed (6.6).

In responding to such questions we shall eventually move away from abstract philosophical issues towards more empirical aspects of science and culture, where sometimes the facts can be allowed to speak for themselves. In the present chapter we continue to explore the domain of individual mental phenomena, looking for evidence in support of our complacent assumption that science is real and true, despite its uncertain origins in the minds of men. Then (Chapter 6) we shall look at the scientific community as a social institution, and the product that it offers to other men.

5.7 *The universality of science*

Let us enquire, for example, whether science as we know it is as *universal* and as *unique* as we often claim. Is there evidence about humankind, about society, or about science itself, that might suggest

---

[30] K. R. Popper (1963) *Conjectures and Refutations* (London: Routledge & Kegan Paul) p. 25. 'I propose to replace the question of the sources of our knowledge by the entirely different question: "How can we hope to detect and eliminate error?"'

[31] A. Naess (1968) in *Scepticism* (London: Routledge & Kegan Paul) makes the case persuasively and sympathetically.

[32] 'We must now recognize belief once more as the source of all knowledge. Tacit assent and intellectual passions, the sharing of an idiom and of a cultural heritage, affiliation to a like-minded community: such are the impulses which shape our vision of the nature of things on which we rely for our mastery of things. No intelligence, however, critical or original, can operate outside such a fiduciary framework'. M. Polanyi, *Personal Knowledge*, p. 266.

The stuff of reality

the contrary? Could our doubts be confirmed by the discovery that science is different in different epochs, in different cultural settings?

Unfortunately, modern science is monolithic and monopolistic. 'Western' science as we know it has unique roots in seventeenth-century Europe and has effectively eliminated all competitors. Apart from its own self-critical activity, it has not been seriously challenged on its own grounds by an independent body of knowledge developed in a quite separate human society. It is well known, of course, that the science and technology of China were comparable with, or superior to those of Europe until the late Renaissance. It is favourable to the thesis of the universality of science that the results of Chinese astronomy are essentially equivalent to those obtained in the West, up till the time of Galileo, apart from such conventions as the definitions of the constellations and the choice of celestial coordinates. In clinical medicine, on the other hand, there are such large differences of categorization and interpretation (7.2) that Western and Chinese doctors might appear to be talking of quite different physiological phenomena.[33] Having conceded that Western medicine is far from complete and consensual, and that it may have as much to learn from Chinese theory and practice as it can pretend to teach, we must note this disturbing evidence that the scientific investigation does not converge inevitably on the same, unique, 'objective' map.

But modern science is unique as a social institution. In no other civilization was there a comparable system of mutually communicating, criticizing and socially-interacting observers (6.3), dedicated to the principles of consensibility and consensuality. What if the Americas had not been discovered for several hundred more years, and the Aztecs or Mayans had made their own 'scientific revolution'; would their knowledge of physics, chemistry and biology have turned out to be equivalent to those of, say, nineteenth-century Europe? Would they have arrived at the laws of conservation of mass and energy, at the theory of evolution by natural selection, at what we call 'Newtonian dynamics' and 'Mendelian genetics'? This comical experiment did not, apparently, suggest itself to the good Lord, so that we do not know the answer; but the question arises, acutely if still hypothetically, in every discussion of the possibility of intelligent life on other planets.

Until the arrival of the little green men in their flying saucers, we cannot advance empirical evidence on this vital point. But we can study several subsidiary questions that bear upon the issue. By looking at

[33] J. Needham, 'The evolution of oecumenical science: the rôles of Europe and China', (1976) *Interdisciplinary Science Reviews*, **1** (1976), 202–13.

existing cultural differences, especially in the structure of *language* we can estimate the extent of possible ambiguity or variation in the perception of the material domain or the communication of knowledge amongst human beings in our own world. By studying the psychological development of the human individual we can get some idea of the powers and limitations of the perceiving/communicating instrument through which all our knowledge flows. Moving outside of science itself, we look for the actual consensual elements in the individual view of material reality, and for the psychological origins of our confidence that there is indeed a domain of being that is external to our own tiny minds.

5.8   *Natural language*

Ideally, scientific knowledge is stored and communicated in a variety of more or less artificial and formalized languages (2.1), or as reproducible material maps (4.1). But the doorway between the public noetic domain and the private mental domain of each individual is open in the first instance only to messages expressed in the *natural language* of his particular social group. A vast amount of scientific knowledge is, indeed, conveyed through these media; even the precise concepts and symbols of mathematics are primarily defined in simple verbal terms. The power, the authority, the credibility of science depend on the characteristics of the medium that links the scientist with his colleagues and puts him in touch with their contributions to public knowledge. And to understand the way in which a person becomes a 'scientist' and learns to 'think scientifically' we have to go right back to infancy, and watch the development and internalization of language in the growing child.

This is a highly controversial topic in psychology, physiology, sociology and linguistics. Crude experimental intervention (7.4) is forbidden, and the interpretation of subtle observations is seldom uniquely convincing (7.5). It is known, however,[34] that before an infant can verbalize it is capable of *sensorimotor coordination*: for example, a perceived object can be deliberately grasped with the hands. This degree of coordination between perception and action – involving, no doubt, many complex transformational processes on the retinal signals within the brain (5.1) – is found in all the higher animals, up to the sophisticated level of problem-solving (5.4) that astonishes us in the dolphin or chimpanzee.

[34] See e.g. Piaget and Inhelder, *Experimental Psychology; its scope and method:* VII.

In the healthy human infant, the neurophysiological machinery for learning languages is innate, and receptive over a certain span of years, although this capacity is developed only by contact with others, by imitation, reward and reinforcement of appropriate responses, etc. Any normal child put into any human cultural group can thus learn to speak 'perfectly' the language of that group. The actual forms of a language (phonemes, words, grammatical usages, etc., etc.) are not therefore personally biogenetic (however well they may eventually be imprinted in the brain mechanisms of the individual person) but are maintained and transmitted as a social institution within the corresponding cultural group. It is interesting that chimpanzees can learn human sign languages[35] and in the wild seem to use a multitude of gestures to convey information beyond individual mood. Contact with the noetic domain does not, therefore, depend on the ability to voice intelligible speech, but is established at a deeper level within the mind.

The process of learning to speak goes on simultaneously with sensorimotor action, which is associated with, guided by, and guides language. Development in infancy towards basic mastery of a spoken language is not at all like learning a foreign language by memorizing a vocabulary and practising the rules of a new grammar. In the first instance, for the child, words and grammatical relations refer to themes that are being continually acted out in the world of sensible phenomena.[36]

Spoken language, in its turn, aids and molds action. And unvoiced words, rehearsed only in the mind, but associated with particular images or actions, are the raw material of conscious thought. 'Thought undergoes many changes as it turns into speech. It does not merely find its expression in speech: it finds its reality and form'.[37] Through primary perception and sensorimotor action, the mental domain (5.5) undoubtedly acquires an operational representation of the material world. But the 'higher' phenomenon of consciousness (with which we are all familiar by introspection) is a consequence of the internalization of language, whose social character draws us irresistibly into the noetic domain.[38]

From this it follows that in acquiring a language a child is automatically made into an instrument of its social group – a device that

[35] R. A. and B. T. Gardner (1969) *Science*, **165**, 664–72.
[36] See e.g. R. Brown (1970) *Psycholinguistics* (New York: Free Press) p. 223.
[37] L. S. Vygotsky (1962) *Thought and Language* (Cambridge Massachusetts: MIT Press).
[38] 'Without language the [logical] operations would remain personal and would consequently not be regulated by interpersonal exchange and cooperation. It is in this dual sense of symbolic condensation and social regulation that language is indispensable to the elaboration of thought.' J. Piaget (1954) *Acta Psychologica*, **10**, 88–98.

ean only transmit messages in the language in which it is brought up, and that could be sensitive only to the categories implicit in that language.[39] Since natural languages are imprecise, ambivalent and formally incomplete, this process of socialization need not be supposed to produce identical, rigidly oriented and blinkered minds (7.10). But it does raise the very disturbing possibility that the noetic domains of social groups with different languages might be so differently organized that the notion of a universal consensual science could not be sustained.

Against this manifestation of 'cultural relativism' there is the evidence from linguistics[40] that all languages conform to the same deeply-lying *transformational grammar* – that any statement that has meaning in any language can be transformed systematically without loss of meaning into a symbolically standardized form (5.4). The child, moreover, in learning a language soon acquires the capacity to take a meaningless statement such as 'The green ideals eat unutterable committees' and turn it into a syntactical equivalent such as 'The committees that green ideals eat are unable to utter': it is as if we were born with the ability to accept the messages of a natural language and transform them logically with the same facility as the central processing unit of a computer can accept and manipulate the coded messages that come to it in the course of any programmed computation, regardless of their arithmetical or physical significance.[41]

Is there any direct neurophysiological evidence for such a specific 'faculty' of the mind? Various neural mechanisms involved in language can be approximately localized in various functional areas of the adult brain, although these seem to have 'moved about' as the child grows to maturity. But these functional areas are certainly not analogous to pre-wired 'black boxes' serving as typical elements in an electronic communication network (5.1). The most significant point from neurophysiology is the close connection between the 'verbal' and 'perception' areas of the brain so that 'cases of complete absence of visual images evocable by words denoting objects are unknown under normal conditions'.[42]

---

[39] 'a mother in expanding speech may be teaching more than grammar; she may be teaching something like a world view', R. Brown, *Psycho-linguistics*, p. 98.

[40] See e.g. J. Lyons, *Chomsky* (1970: London: Fontana/Collins).

[41] 'many of the universal features of language, both substantive and formal, are not readily explained... The only conceivable explanation, says Chomsky, in terms of our present knowledge at least, is that human beings are genetically endowed with a highly specific "language faculty" and that it is this "faculty" which determines [these]... universal features'. Lyons, *Chomsky*, p. 105.

[42] A. R. Luria, *The Working Brain*, p. 273.

Now we have evidence from child psychology that visual transformation comes before verbal manipulation in the development of the brain. The primacy of the unverbalized sensorimotor scheme – i.e. the mental map – is indicated by the fact that blind children who can speak find difficulty in forming such schemes, whereas deaf children who have great difficulties with words as such can 'visualize' and solve problems without the aid of verbalization.[43]

Returning to the processes by which language is normally learned, we see that these would be gravely impeded if the structure of language were not reasonably close to the structure of the sensorimotor action with which it is invariably associated at this stage. But 'sensorimotor coordination' can be functionally equivalent to a system of logical relations:[44] thus, for example, the operational world of the child contains 'graspable', more or less permanent, 'objects' whose interrelations make actual the logic (or grammar!) of existence, number, equivalence, identity, etc.[45] A language that did not directly symbolize these relations would be unintelligible and unacceptable to the child. Although, in principle, a social group might cultivate such a language, and impose it on the captive audience of its children, it is hard to imagine how it could have evolved, or could maintain itself under natural conditions, against the need to be learnt anew in each generation.

On this view point, the 'logicality' of natural language and of conscious thought is neither imposed arbitrarily from society in the preformed structures of a possibly exotic language, nor is it necessarily 'wired-in' genetically in the brain: it is necessitated by the behaviour of things 'out there' in the material domain, which every individual experiences as he grows and learns.[46]

Piaget's theoretical scheme does not particularly depend on the child being born with a 'language faculty' which is pre-programmed to accept and process the grammatical code of a natural language. There

---

[43] H. Sinclair-de-Zwart (1969) in *Studies in Cognitive Development* edited by Elkind and Flavell (Oxford: University Press).

[44] J. Piaget (1954) *Acta Psychologica*, **10**, 88–98.

[45] 'the origin of intellectual operations is to be sought in the subject's actions and in the experiences and lessons arising out of them. These experiences serve to show that such actions, in their most general coordinations, are always applicable to the object; e.g. [a baby] uses the order of his actions of exploration to establish the order of the objects.' Piaget and Inhelder, *Experimental Psychology; its scope and method:* VII.

[46] 'Operational structures i.e. logic are more closely linked with coordination of actions than their own verbalization'...'operations are in origin neither social nor individual in the exclusive sense of the term. Rather do they express the most general coordinations of actions, whether they are carried out in common or in the course of individual adaptations' Piaget and Inhelder, *Experimental Psychology; its scope and method:* VII.

is direct evidence, however, that the learning of a language is not passive imitation and adaptation in the behaviourist mode.[47] The child spontaneously generates grammatical transformations of increasing complexity, as if the innate brain mechanisms postulated by Chomsky were being activated and brought into use. How does this apparently autonomous faculty of the mind fit into the scheme?

This is a matter for speculation; the following remarks are entirely conjectural, if not entirely irrelevant to our general theme. The hypothesis is that the 'language faculty' is not to be found in a completely distinct 'verbal processing unit', but is largely associated with the visual perception mechanism of the brain. As we have noted, this mechanism develops rapidly in early infancy, and is well formed by the time that the child is ready to learn a language. The organs provided by anatomical morphogenesis are shaped and articulated by sensorimotor experience into a neural network that can respond efficiently to what is to be perceived. The characteristic structural relations of the material domain – object, identity, similarity, permanence, spatial topology, dimensionality, motion, time sequence, etc. – must have their counterparts in the relationships between the corresponding patterns of neural excitation. We do not know what these patterns are, but we know some of the rules they must obey. Within the brain there must be 'mappings' of characteristic features and aspects of the everyday world that conform to the inherent logic of action in that world – for example, neurones associated together in the actual topology of an observed visual pattern.

The language that is now to be learnt must be structurally consistent with these mappings, to whose features it will inevitably attach itself by verbal association. These 'features' may themselves be relational, for we know that the brain can 'abstract' such characteristics as connectivity and motion from the image on the retina, and does not merely record catalogues of 'objects' or primary sense data (5.1). In the linguistically mature mind, the word creates image-like excitations in the visual regions of the brain, whilst images drawn from perception or introspection give rise to verbalizations in conscious thought.

It must be assumed that a verbal statement excites a map-like image conformal with its logical structure. Thus, for example, the assertion that 'all swans are birds' must excite an image that features the set-theoretical relation of 'inclusion' – perhaps, quite simply a Venn diagram (5.4) showing a spatial region labelled 'swans' entirely enclosed within a larger region labelled 'birds'. Some such structural

[47] R. Brown, *Psycho-linguistics*, pp. 100–54.

isomorphism is essential if there is to be a meaningful transfer of information from the seeing eye to the speaking tongue.

From our study of maps and pictures (4.4) we have become familiar with the image-processing faculty that underlies pattern recognition and visual perception. Introspection of the mental processes that take place, for example, when we correctly assign an empty glove to the left or right hand reveals that we can mentally transform a spatial image by lateral displacement, rotation, distortion, etc. without losing cognisance of essential geometrical or topological properties. Is it fanciful to suppose that these map transformations are easily applied to the unconscious images aroused by verbal assertions, generating topologically equivalent images that are 're-read' as different but grammatically or logically equivalent statements? Looking at that internal, unconsciously generated Venn diagram we say it another way: 'Amongst birds there are swans.'

The mechanisms of primary sensorimotor coordination studied by Piaget are certainly powerful and flexible enough to generate the multifarious linguistic forms of Chomsky's transformational grammar. These transformations, moreover, are not chaotic or arbitrary; they must in any case conform to the basic topological structure of the 'visual areas' and hence, eventually, to the 'natural' structural relationships of the material domain. If this speculative interpretation is accepted, we can see good reason, approaching necessity, for the universality of a basic 'commonsense' logicality in all human languages, derived from the universality of childhood experience within a growing human body, surrounded by, nurtured by, our own kin.

5.9 *Cultural variations in cognition*

Faced with the wild contradictions and diversity of the world pictures current in different human cultures (7.10) we might forever despair of finding any consensual basis for science. But these contradictions and diversity are no more serious than those to be found amongst, say, scientific cosmologists at the edge of established knowledge (6.5). We have already found satisfactory universality in the underlying logic of natural languages, permitting at least some approach to consensibility and consensuality. There are adequate means for communicating unambiguous messages between the members of different language groups.

The question has often been asked whether all human beings see the world in the same way – whether they are interchangeable as observers in the scientific system (3.1). By now we can immediately answer this crude question in the negative; the scientific maps and other paradigmatic schemes imposed upon the individual mind out of the noetic domain have a dominant effect upon perception and cognition (4.4). Yet we must continue the search for aspects and elements of the world view that are effectively universal amongst mature men and women.

It is known, for example, that most natural languages assign names to colours according to much the same scheme – in accordance, presumably, with the universality of the physiological mechanisms of rods and cones, retinal pigments and neural receptors in the human eye.[48] A more striking phenomenon is the appearance of the same metaphor – e.g. the verbal–visual synesthesia of a phrase such as 'John is a very cold person' – in languages as unrelated as Hebrew, Greek, Chinese, Thai, Hausa and Burmese.[49] It is hard to believe that this apparently spontaneous manifestation of the same underlying relationship in the mental domains of many different peoples can be entirely coincidental; they must, we suppose, 'feel' such things in the same way (7.9).

On the other hand languages certainly exhibit considerable diversity in their descriptive and metaphorical resources. The Eskimo has thirty different words for snow; the Arab knows innumerable variations of the beast we call simply a camel. Hopi Indians do not use our familiar metaphors for duration and intensity; their language contains no analogues of phrases such as 'a *long* time' or '*widely*-held views', whose meaning is drawn from an imagined space where the passing of time is represented along a spatial dimension and a human group is represented as a spatially dispersed set of elements.[50] But the significance of such observations has been much disputed. Granting the diversity, both in riches and in metaphorical poverty, of the noetic domains enshrined in different cultures with different languages, we must probably agree that 'languages really only differ in what is relatively easy to say'.[51]

Perhaps the most striking example of a distinct cultural variation in primary cognition is reported from New Guinea.[52] Amongst the

[48] 'colour space is a prime example of the influence of underlying perceptual-cognitive factors in the formation and references of linguistic categories' M. Cole and S. Scribner (1974) in *Culture and Thought* (New York: Wiley) p. 55.

[49] Cole and Scribner, *Culture and Thought*, p. 56.

[50] B. Whorf (1941) in *Language, Culture and Personality*, edited by L. Spier (Salt Lake City: University of Utah Press).

[51] Cole and Scribner (*Culture and Thought*, p. 44) quoting Hockett.

[52] In *Socialization: the Approach from Social Anthropology*, edited by A. Forge and P. Mayer (1970: London: Tavistock).

Abelam, in the context of the tambaram cult, boys and young men acquire a set of fixed expectations about what they will see in two dimensions: their polychrome two-dimensional paintings become a closed system, unrelated to natural objects, or to carvings, or three-dimensional art objects – so much so that they cannot 'see' (i.e. make sense of) anything in two dimensions (e.g. photographs) that are not in this system. The 'nggwalndu' paintings do communicate a great deal to the Abelam, but 'holistically' and not in ways that can themselves be paraphrased in speech.

It would be wrong to suppose, however, that the material world perceived by the Abelam is warped (by our standards) beyond recognition in the mental domain of each individual. They behave quite normally towards everyday things – and can be trained in a few hours to interpret photographs! We have here, perhaps, no more than an extreme case of the cultural diversity of artistic traditions and conventions for the material representation of the visual world.[53] It is not necessary to assume that the difference between an Egyptian tomb painting and an eighteenth-century caricature corresponds to a fundamental difference in the internal visual schemes of the respective artists. One must assume that the sensorimotor coordination schemes developed in the brain in childhood are far more complex, multi-dimensional, plastic and transformed by abstraction than any artist could hope to depict. An artistic convention selects and projects features from the internal image (4.3) and in its turn may lay its own structure upon some aspects of our thoughts. But this, surely, is a secondary process, akin to the projection of other socially-held paradigms of concept and language on the developing individual mind.

Unfortunately, the further exploration of such subtleties is effectively blocked by formal *schooling*. In almost all modern societies, the taught written languages conform far more closely to universal standards of content and structure than the spoken languages from which they derive. Written communication is essential to science; but the forms of writing that are taught are deliberately chosen to develop the powers of 'abstract' thought that are characteristic of scientific knowledge.[54] This process of socialization towards the paradigms of the modern technical society (7.10) rapidly obliterates genuine cultural differences in the perception of reality and in the 'powers of human

[53] E. Gombrich, *Art and Illusion*.
[54] 'writing, then, is a training in the use of linguistic contexts that are independent of the immediate referents.' P. Greenfield, L. Reich and R. Olver (1966) *Studies in Cognitive Growth* (New York: Wiley).

thought'. A person brought up in a pre-literate traditional culture may well 'see the world' in interestingly different ways (6.7); but that vision cannot be communicated in the consensible language of science, which would trample on the fragile neural net within which it was conceived.

Indeed, the education required to make a scientist (4.4, 6.2) must have this effect of reducing individual variations of primary cognition within a single culture. To take part in scientific activity, one must be able to communicate efficiently with other scientists. Experiments in which individuals have to describe objects before them for others to identify show that uneducated adults often fail because they do not consider the information that the *other person* needs.[55] This brings out clearly the way in which language as a means of 'communication' develops and needs an element of 'communion'; in becoming more 'intersubjective' it has to become more 'objective' (5.6), hence leaving little room for personal variations. In this pursuit of scientific precision, we cultivate the practice of *maximum message communication*[56] which spells out all the details and imposes a prosaic consensus in the interpretation of the material domain. In Japan, by contrast, the tradition of *minimum message communication* assumes a pre-existent empathy or cognitive consensus that need not be further reinforced. It is possible that this freedom for 'poetic licence', even in material matters, explains the poor development of logic in Japan, where science did not arise spontaneously, despite the high rate of literacy.

5.10   How much is real?

Strict application of the principles of the sociology of knowledge seems to lead to the inescapable conclusion that science is no more than one of many competing world pictures in the noetic domain,[57] and is not

[55] Cole and Scribner (*Culture and Thought*, p. 180) also quote Piaget 'it is only after long training that the child reaches the point...where he speaks no longer for himself, but from the point of view of the other'.

[56] N. Nagashima (1973) in *Modes of Thought* edited by R. Horton and R. Finnegan (London: Faber). Cf. also the distinction between the restricted and elaborated speech codes used to convey mythical narrative and 'scientific' discourse respectively: see B. Bernstein (1971–73) *Class, Codes and Control*, I–II (London: Routledge & Kegan Paul). The issues underlying this section are discussed very clearly by V. E. Turchin (1977) in *The Phenomenon of Science* (New York: University of Columbia Press) Chapters 7–8.

[57] B. Barnes (1974) *Scientific Knowledge and Sociological Theory* (London: Routledge & Kegan Paul) p. 42. 'As to normal patterns of belief and action, these must, in the first instance be treated as culturally *given*. They are maintained as permanent institutional features through socialization and social control, and it is technically impossible to explain them via their ultimate origins...We may claim that all institutionalized systems of natural belief must be treated as equivalent for sociological purposes.'

privileged by comparison with any other systematic scheme to which a social group can subscribe, such as the famous magical beliefs of Azande. But total *cultural relativism*, like complete philosophical scepticism (5.6), is a sterile doctrine that inhibits further interesting and valuable investigations.

To escape from it, let us avoid the vulgar error that the essence of science lies in its most sophisticated results, its astonishing revelations of the nature of things, and the attendant metaphysical interpretations which contrast so dramatically with alternative, 'non-scientific' world pictures.[58] It is sheer intellectual snobbery to fasten solely on those high-flown features of modern scientific knowledge, ignoring the mundane foundations on which they rest.[59] Our model of science emphasizes the whole corpus of consensual knowledge, which is not necessarily true in every detail (6.5), which may yet contain gross conceptual errors and fallacies (4.5), but which is not to be judged simply by inspection of its most extreme and schematized theoretical consequences (4.4).

The available evidence (5.8, 5.9) gives us no strong reason to doubt that there is nearly universal human consensus concerning certain aspects of the material domain. All mankind, through its natural languages, exhibits its adherence to the elementary principles of logic (2.3), and through its senses discovers a mentally coherent world of invariant objects and spatial relations, patterns of sound and colour, permanence and movement, time and change. This *matter-of-fact world*, projected into the noetic domain by language, is nearly the same for us all. But the universality of *commonsense realism* is not due to all human minds having been genetically created as identical perceiving/communicating instruments like remotely controlled television cameras landed on Mars. We derive our well-founded confidence in the reality of everyday things from the sensorimotor bodily experience of childhood, reinforced and brought to consciousness by linguistic interaction with other people (5.8). If we are to make any sense of science at all, we must first agree that the sun is hot, and the moon

[58] I.e. emphasizing the 'characteristically abstract and theoretical nature of scientific knowledge, which sets it apart from common sense or the more concrete lore of the chef or builder'. Barnes, *Scientific Knowledge and Sociological Theory*, p. 46.

[59] This tendency has been nicely characterized by D. Bloor (*Knowledge and Social Imagery*, p. 42). 'Sociologically speaking knowledge has its sacred aspects and its profane side, like human nature itself. Its sacred aspects represent whatever we deem to be highest in it. These may be its central principles and methods...the sacred aspects of science can be thought of as informing or guiding the more mundane, less inspired, less vital parts. These are its routines, its mere applications, its settled external forms of technique and method.'

is bright, that stones sink and wood floats, that he who falls from a tree may break his bones, and she who does not eat will die from hunger. On all such matters the 'savage mind' is no worse informed, no more confused, than the most brilliant product of the Cambridge Natural Sciences Tripos, with his knowledge of quantum theory, relativity and the structure of DNA.[60]

The study of natural languages, of psychological development and cultural differences of perception and cognition, thus assures us of a region of consensibility, of an unambiguous, universal, *categorial framework*[61] covering at least part of the external world of nature. This region might be described as including anything that could be the subject of a good conjuring trick: a rabbit 'materializing' from an apparently empty hat; a living woman being 'sawn in half'; a 'smashed' watch perfectly reconstituted. The stage magician makes his living by creating the illusion that one of the elementary principles of common-sense realism has been contradicted: throughout the world, in every human culture, the same tricks are effective as sources of amazement and curiosity. In some societies, of course, the illusionist does not reveal his methods, and imposes himself as the possessor of supernatural powers (6.6). But the credulous behaviour of his audience – whether genuine or feigned – is a comment on the *social* role of the magician: he is a *superhuman*, who can interfere with the ordinary everyday way of things, whose uniformity is otherwise fully accepted. The concept of a *miracle* implies strong belief in an agreed natural order, from which deviations can be caused only by the intervention of a divine personality outside of nature.

It must be emphasized, indeed, that the universal consensus concerning certain aspects of the physical world does not extend to the whole noetic domain. *Psychological* and *social* knowledge are not acquired in early infancy, and are subject to the full variability of strong cultural relativism (7.10). For the moment, however, we are questioning the credibility of scientific knowledge at its best, in the physical and biological sciences, where we feel sure that it can be relied on with considerable confidence. All our doubts concerning the applicability of the same methods to the study of human behaviour will be fully aired in Chapter 7.

On the other hand, the categories and logical relations of the

[60] The reasons why anthropologists have tended to emphasize the *differences* between savage religion and civilized science are explained by R. Horton in *Modes of Thought* (p. 263) edited by R. Horton and R. Finnegan (1973: London: Faber).
[61] S. Körner (1970) *Categorial Frameworks* (Oxford: Blackwell).

everyday world, although continuously grounded on human experience and refined to self-consistency by the social institution of language, are not restricted to what can be expressed symbolically, or in quantitative terms, according to the finite algorithms of mathematical representation (Chapter 2). Commonsense discerns in the material domain a variety of extra-logical, non-verbal, yet highly consensible aspects of nature – characteristic patterns, as of biological species (7.2); adjacencies and sequences, as of geological strata; connections and interactions, as of many mechanical devices. Everyday reality is far more widely *objectively pictured* (4.3) than it has been, or can be, objectively *mapped* or *described*.

A question that must remain open is whether the categorial framework of everyday realism is the only possible scheme for describing or understanding nature in its physical aspects. The predominance of two-valued logic (2.6) in the deep grammatical structure of all natural languages is very convenient in practical affairs, but does not seem to be absolutely prescribed. It is a nice question whether a community of intelligent jellyfish, or 'Daleks', or 'Black Clouds',[62] living in conditions where sharply defined, independently invariant objects were not familiar, would necessarily develop a language based upon the same logical principles. A language based upon, say, a standard three-valued logic ('yes', 'no', 'maybe') would seem to us highly contrived and artificial, but would not seem much more difficult to learn or use in daily life if we really needed it. A science communicated in that language would perhaps be more reliable in its firm conclusions, more honestly uncertain, than the crudely chopped material that we now tend to get. Or would all our conjectures, all our data, all our predictions, dissolve and fade in doubt and scepticism, offering no more guidance to action than the paradoxical poetry of a Zen *Koan*.

This chapter, devoted to various aspects of individual *cognition*, thus ends inconclusively. This is as it should be. The mind of no man is a deterministic machine. The behaviouristic robot is a myth, of value only as a primitive 'model' in a few trivial psychological investigations. In attempting to talk about perception and language, consciousness and belief, we expose our argument to quite incalculable influences. These are mysteries, ruled by disconcerting paradoxes and wild ignorance. If the sciences of psychology and neurophysiology can arrive at better understanding than, say, the ancient art of introspection, then good luck to them: but the obstacles they face are not imaginary (7.10) and we can have no certainty that they will eventually be overcome.

[62] F. Hoyle (1973) *The Black Cloud* (New York: New American Library).

Meanwhile, we must accept the fact that all our beliefs about science are subject to just these mysteries. Neither the logic of unambiguous communication, nor mechanical instrumentation, nor ideal norms of scientific research can save us from uncertainty or error in the way we picture the world about us. But what would there be for us in science, if we could not live as dangerously there as in other realms of being?

6

The world of science

'Not the testimony of all the Fellows of the Royal Society, nor even the evidence of my own senses, would lead me to the belief in the transmission of thought independently of the recognized channels of sense.'

Helmholtz.

6.1 Specialization and authority

The whole strategy of science is directed towards the creation of a maximum consensus in the public domain. Such a consensus must be based on, and held together by, a pre-existing mental harmony[1] between independent human beings on at least some matters of common interest. For the natural sciences, this consensual basis is immediately available in the everyday world of which every child early becomes aware through experience and the social medium of natural language (5.10).

Unless he has been got at by philosophy, the average bench scientist is quite prepared to swear that his branch of science is just *commonsense*. He will agree, of course, that he himself may be an extremely *un-common* type of person, with a very specialized education; yet he does not draw any mental distinction between his scientific knowledge and his practical acquaintance with everyday things, and would bluffly assert that his scientific modes of looking for evidence and arguing towards inferences are not different in principle from what he would do if he had to mend his motorcycle or detect a murderer.[2] And he does not regard the electrons, or amino acids, or genes, or extinct

---

[1] 'What guarantees the objectivity of the world in which we live is that this world is common to us with other thinking beings. Through the communication that we have with other men we receive from them ready-made harmonious reasonings. We know that these reasonings do not come from us and at the same time we recognize in them, *because of their harmony*, the work of reasonable beings like ourselves. And as these reasonings appear to fit the world of our sensations, we think we may infer that these reasonable beings have the same thing as we; thus it is that we know we haven't been dreaming. It is this harmony, this quality if you will, that is the sole basis for the only reality we can ever know': R. M. Pirsig (1974) in *Zen and the Art of Motorcycle Maintenance* (London: Bodley Head) p. 262.

[2] Indeed, he will probably add that scientific ways of thinking are precisely those most likely to be effective in practical affairs, and that the everyday world itself would be a better place if everybody adopted a 'scientific attitude'!

hominids with which his research is concerned as any less 'real' than the cakes of soap in his bathroom or his own bonny children.

Such an attitude is an affront to the intelligence of the ordinary person. He knows that scientists use very complicated apparatus, and communicate in mathematical symbols. He has been told, in all seriousness, that a table is 'really' a buzzing swarm of electrons and nuclei, that space is permeated by immense currents of enormously energetic neutrinos that can scarcely ever be observed, that all life is merely the mode of self-replication of DNA, and similar marvels (4.4). The layman accepts what the scientist tells him in the same spirit of wondering credulity as he formerly accepted the theological speculations of the priest – but he is not such a fool as to confound the 'mysterious universe' revealed by science with his own homely world. In other words, he credits the scientist with access to ways of thought which are fundamentally different from his own; he is persuaded to believe in science, not by the weight of the evidence in the scientific archives but in the light of the intellectual *authority* of the scientific expert.[3]

The equivalent observers (3.1) of our schematic model of science (1.3) were assumed to be any ordinary human beings. The intellectual consensus that might eventually be hammered out inside such a group could then be extended to the whole of humanity, and thence, we must suppose, to all rational beings. In these idealized circumstances the credibility of science would be seen to depend upon the diversity of unbiassed attitudes amongst the members of this 'jury', who would critically assess all the evidence and listen patiently to every side of the case.

In practice, however, scientific knowledge is generated and validated by a *scientific community*, which is as far as could be from a random sample of unregenerate mankind. By the social division of labour, modern society entrusts the cultivation of science to a highly specialized professional group, characterized both by expertise and extreme commitment to science as a social institution. What should, in philosophical principle, be done by all men, is given into the hands of proxies, who bear collectively the powers and the responsibilities of science within society at large.

This specialization and professionalism in the practice of science is

---

[3] If we accept, with Popper (*Objective Knowledge*, p. 34) that 'All science and all philosophy are enlightened common sense', we must admit that some forms of sense are not so common after all, and some people are much more enlightened than others.

immensely effective for the production of new knowledge and its application. Let us not doubt the sociological necessity, the historical inevitability, of this development. The vast extent of scientific knowledge, and its intellectual depth at every point, are beyond the critical assessment or re-assessment of the uninstructed layman. But it does mean that the general criterion of the *universality* of science (5.7) is not to be taken as a serious demand for an active consensus of all mankind. Although we may assert in principle that science is potentially persuasive for any rational being, this is a practical impossibility. The scientific community is assigned the duty of acting on behalf of all citizens in the creation and criticism of public knowledge, in the reasonable expectation that its judgements will not be challenged.

In assessing the reliability of the science thus produced, we cannot avoid asking how well this specialized social institution actually performs its assigned role. We know, for example, that even the most dedicated and incorruptible human group is not immune from collective fallacies and mass delusions (4.5). The *quality* of the knowledge amassed in the scientific archives depends much less upon philosophical paradoxes or psychological quirks than upon the training given to individual scientists (4.4), their mutual relationships in the scholarly network, and the place of science in society at large. The *sociology* of science is relevant to our discussion, not simply because science is a highly active transforming factor in society, but because the cognitive contents of science depend for their form and integrity on the manner by which this social institution shapes and governs its members.

6.2 *Learning science*

The would-be scientist must first learn his 'subject'. It is not enough to have technical expertise in skills such as algebraic manipulation or electronic circuitry; it is necessary to be fully acquainted with the conceptual foundations of current research, and to grasp the contemporary paradigms of a discipline (4.4).

Learning to 'think scientifically' (i.e. as a physicist, or a chemist, or a palaentologist) is a long and complex process. On the one hand the student cannot simply learn science by 'personal discovery'. Faced with an apparently meaningless collection of apparatus and phenomena he is quite incompetent to reproduce the scientific steps of innumerable predecessors by his own unaided efforts. Scientific concepts do not spring up out of experimental facts, any more than three-dimensional geometry synthesizes itself automatically out of sense data, by induction and inference.[4] It is impossible to acquire a grasp of the sophisticated languages and patterns of scientific thought without firm guidance from a fully qualified teacher or from books that expound the current consensus.

On the other hand, the learning process cannot be speeded up by teaching scientific facts and theories by rote, to be memorized in bulk as one might the vocabulary of a foreign language, or the map of some distant country. It is not merely that *indoctrination*, in the authoritarian tradition of theological or ideological education, is antipathetic to the criticism and scepticism (5.6) that are essential to the research profession. It is that scientific concepts only become *real* by practical use.

To the philosopher, science is interesting in its abstract theories; to the person in the street, it is valued for its practical achievements; but to the scientist it is the *unity* of theory and practice that he most prizes, and emphasizes in his teaching. In almost all the disciplines of the 'Natural Sciences', there is a strong tradition of training in experimental and observational techniques – repeating famous historical experiments, using test-tubes and microscopes, observing rock formations and identifying minerals in the field. In mathematics and other symbolic disciplines, every student must show that he can use the formal theories to solve 'examples' and applied problems. Many science teachers regard this sort of training as the only means of acquiring proficiency in the technical arts associated with research[5]: it is even more significant as the source of personal *experience* which combines with socially institutionalized *theory* to produce that sense of *reality* which every scientist feels about his own subject (4.4).

[4] J. Nicod, *Geometry and Induction*, is a fascinating attempt to do this. In particular the importance of going beyond the individual viewpoint becomes apparent. 'Now what establishes geometric order among the views apprehended by the various subjects, each of which has his personal space, is not yet the perfect resemblance of these spaces, but the systematic *differences* of the perceptions they have as contents. In order for the views of all to be coordinated in a single spatial order, it is not sufficient that they have *a priori* similar structures: it is in addition necessary for them to have different sensible contents – different according to the complicated and precise rules of perspective. It is these differences between your perception and mine, each in its own immediate space, that constitute the single space in which both of us are situated' (p. 131). Eventually this 'space of views' becomes the primary geometry. Nicod's analysis thus foreshadows the significance that Piaget assigns (5.9) to 'seeing through the eyes of others' in the construction of reality.

[5] 'The large amounts of time spent by students of chemistry, biology and medicine in their practical courses shows how greatly these sciences rely on the transmission of skills and connoisseurship from master to apprentice. It offers an impressive demonstration of the extent to which the art of knowing has remained unspecifiable at the very heart of science.' Polanyi, *Personal Knowledge*, p. 55.

In other words, the scientist learns to talk and think scientifically in the same way as the infant learns to talk and think about the world of everyday reality (5.8). On the one hand, the science student, like the baby, practices by personal manipulation, in experiment or in symbolic operations, until he knows the natural logic, the 'sensori-motor coordination' of the scientific objects and concepts with which he is provided: on the other hand, he is kept in touch with the noetic domain, being continually fed, through the social medium of language, lessons, lectures and books, with the public analysis or map of his experiences.

In this process, the infant has the advantage that his sensorimotor experience comes prior to its linguistic representation. The science student usually encounters a new scientific entity such as a 'magnetic field' first as a theoretical abstraction, and has then to manipulate magnets and iron filings until it becomes real to him. In due course, he may be hustled and bullied into professing to see it as part of an 'electromagnetic field' which may then be dissolved mysteriously into a gas of 'photons', or warped geometrically into the 'off-diagonal components of a skew tensor in space-time'. In his brief under-graduate training he seldom has time or opportunity to internalize the whole paradigm in all its richness and diversity, and may leave the university with little better than an uncertain indoctrination in the more advanced aspects of his subject.

Given a good education, and adequate research experience the scientist can picture the scientific map (4.4) into reality. Although it originated outside himself, as a social construction, it becomes as personal and individual as his own backyard.[6] The layman, faced with the incomprehensible subtleties and complexities of a well-established body of science, is ready to believe in it as in a marvellous myth, a mysterious epic whose characters are, no doubt, historically founded, but unreal as people. For the scientist, on the other hand, these characters acquire the solidity and many-sidedness of personal friends; he recalls the exploits they have been in together, and looks forward to travelling with them into unknown regions.[7]

At the outset, the consensual formulations of science rest firmly

[6] 'The field of so-called knowledge is that of the assimilation of the experience of others to one's own experience' G. H. Mead, *The Philosophy of the Act*, p. 50.

[7] 'the scientist, when he times microscopic oil drops, or when he measures the distance of photographed stars from one another before or during an eclipse, has not the attitude of a man perched insecurely upon obscure and adventitious data. The world that is there has taken upon itself all the order, definition and necessity of earlier scientific advance' G. H. Mead, *The Philosophy of the Act*, p. 49.

enough on the universal human cognition of a real world of everyday things (5.10). This is the starting point of all scientific education. But the specialized training of the would be scientist rapidly introduces him to other aspects of nature where elementary 'commonsense' is an inadequate guide. By educational procedures acting precisely on the psychological mechanisms that opened the everyday world to the infant – coherent action, non-contradictory communication, verifiable predictions, consensual cognition and cogitation – he is made to see the 'scientific world' in a whole album of maps and pictures drawn from the scientific archives. For him, that world is not foreign, mysterious, uncertain, unreal; he assimilates it into his own mental domain as a mere elaboration and extension of the commonsense world of every-day things which he shares with all mankind.[8]

This form of enlightened, pragmatic, indoctrination is almost essen-tial if there is to be progress in science. The newly educated scientist must be ready to make *new* discoveries, and explore *new* regions of nature; he cannot spend years of his life retracing and testing all the old maps to be quite sure that he is not being misled. It would be asking too much of him to imagine his way back into the successive levels of ignorance of his forebears, and then to follow each painful forward step of observation or interpretation. Nor is it feasible to mount a complete curriculum of experiments and theoretical workings to cover all aspects of a subject, meet every potential criticism and to satisfy every reasonable doubt of the most sceptical student. It is impossible to learn a science without having a good deal of faith in the competence and sincerity of one's teachers, and accepting their word for what does not seem altogether obvious or true.

For this reason, the question of the reliability of a particular branch of science cannot be settled solely on the basis of the personal opinions of individual scientists within that discipline. The educational process is too sketchy; it is too easy to be indoctrinated (4.4) into giving the attributes of reality to a socially conformist fallacy (4.5). Organized scepticism and competitive criticism are the norms of the scientific

---

[8] A. Schutz (1967) *Collected Papers* I, *The Problem of Social Reality* (The Hague: Martinus Nijhoff: first published 1945) assigns the world of science to a different 'finite province of meaning' from the 'everyday world' in whose paramount reality we are induced to believe 'because our practical experience prove the unity and congruity of the world of working as valid and the hypothesis of its reality as irrefutable. Even more, this reality seems to us to be the natural one, and we are not ready to abandon our attitude towards it without having experienced a specific *shock* which compels us to break through the limits of this "finite" province of meaning and to shift the accent of reality to another one'. In my opinion, scientific education is designed to avoid giving any such shock to the scientist, who wants desperately to unite all worlds under the scientific banner.

community (1.4, 2.9, 3.3, 6.3) but most creative scientists are personally optimistic about the current consensus in their field and overestimate the permanence of what they believe to be already firmly established (6.5). In this respect, also, the official teachings of educational institutions do not give satisfactory guidance as to what is truly worthy of scientific belief.

6.3 Dissent and selection

The intellectual authority of science does not lie in the technical expertise of its individual practitioners nor in the vast archives (6.5) to which their learning gives them access; it resides in the processes by which scientific knowledge is created and accredited. The sources of its reliability are not simply that it is expressed in unambiguous language and is capable of experimental verification; they are to be found in the historical processes of its growth, and in the social relationships of those who brought it into being.

By 'jobbing backwards', it is easy to misrepresent the history of science as a tale of unqualified success, in which one enlightened genius after another was driven by the logic of his situation to make the inevitable step forward. No doubt there is something to be learnt about 'method' in research and invention by studying only successful achievements. But such optimal paths are historical fictions, arrived at only by ignoring a vast body of other material that was no less scientific in its time.

The true record of scientific development includes a great amount of *unsuccessful* experimentation and theorizing that is later kindly forgotten (2.10). This is a fact that can easily be verified by reference to any comprehensive, bibliographically conscientious history or review article, whether reporting on the grander issues of the search for the fundamental laws of Nature, or concerned only with some minor question in a particular discipline, such as the cause of lightning or how much grass a cow eats in a day. *Only a small proportion of the information contributed to science by research is eventually incorporated permanently in the body of scientific knowledge.*[9]

But this is implicit in our model of science (1.3). The network of

[9] 'about one-half of all papers that are published in the more than 2100 source journals abstracted in the *Science Citation Index* do not receive a single citation during the year after the paper is published'. J. R. Cole and S. Cole (1973) *Social Stratification in Science* (Chicago: University of Chicago Press) p. 228. H. W. Menard (1971) in *Science: Growth & Change* (Cambridge, Massachusetts: Harvard) p. 98 remarks that the *Bulletin of the Geological Society of America* (1888–1969) did not cite at all the work of about 80% of the names listed in the *Bibliography of North American Geology*(1785–1860).

scientists is not merely an extended observational apparatus; it is an instrument which analyses and selects for preservation only those messages that receive overwhelming consensual support. In addition to the theories (2.5) that establish themselves by confirmed predictions (2.9), there are numerous hypotheses, conjectures, unreproducible observations and erroneous discoveries (3.6) that have been reported by individual scientists but are not validated by the community. Out of a medley of imperfectly expressed, inconsistently argued, inexplicable and unexpected messages, only what is fully convincing is retained. Drawn towards the unattainable goal of a complete consensus, scientific knowledge *evolves by critical selection.*[10]

In emphasizing this principle, we draw attention to two contrasting features of science as a social institution: *tolerance of dissent* and *critical evaluation* of all contributions to knowledge. It is a matter of debate[11] whether these characteristics of the 'open society' are to be found in primitive human culture or other civilizations or whether they were uniquely combined in European society after the Renaissance: the fact remains that they are the constitutive principles around which the modern scientific community now functions.[12]

On the one hand, the ethos of science calls for creativity, i.e. the production of *novel* ideas that may conflict with received opinion. Contradiction and dissent are not merely tolerated; they are warmly welcomed and, if successful, richly rewarded (3.3). Any speculative hypothesis with a plausible promise of explaining a mysterious phenomenon is given a hearing – especially in fields such as cosmology and high-energy physics where our knowledge borders on regions of complete ignorance. Inconsistent interpretations of the same data are often published in the same number of the same journal, and logically incompatible models (2.9) often co-exist in the literature.

Experienced scientists know that real progress in research is slow and painful, and that it is often better to publish an interesting and novel idea, however incompletely explored and inadequately comprehended, than to keep it private until all its implications have been understood. If scientific knowledge is to evolve, it needs more than the accumulation of new 'facts' enlivened by occasional accidental

---

[10] Many of the greatest names of science and philosophy have gone on record in support of this general principle of 'Evolutionary Epistemology'. The history and larger intellectual context of this principle are admirably summarized by D. T. Campbell in contributions to *The Philosophy of Karl Popper* edited by P. A. Schilpp (1974: LaSalle, Illinois: Open Court Publishing Co) pp. 413–63 and *Studies in the Philosophy of Biology* edited by F. J. Ayala and T. Dobzhansky (1974: New York: Macmillan) pp. 139–61.

[11] See e.g. E. Gellner (1973) in *Modes of Thought*.

[12] See e.g. Ziman, *Public Knowledge.*

discoveries; there must be a public supply of suggestions, hints, possible models, heuristic formulae, fruitful analogies and other components of theory, to stimulate new formulations and 're-mappings' of the existing observational data. In order that science may continually break through the invisible barriers of its own paradigmatic categories, each scientist is encouraged to be an imaginative source of interpretation, both of his own contributions and of the work of other scientists.

On the other hand, nothing may be published as scientific information without careful, critical scrutiny by editors, referees and reviewers. The highest standards of instrumental accuracy and logical necessity are imposed on all scientific communications. Experiments are conscientiously repeated (3.5) and theoretical calculations tested by alternative procedures. Every scientific paper, ostensibly building on the preceding work that it cites, carries an implied or open criticism of much of that work, which it seeks to validate or disconfirm and supersede. Review articles, colloquia and research monographs delineate controversial issues, and delicately point out the deficiencies of many reputable research contributions.

Experienced scientists know, indeed, that real progress in research is slow and painful, and that many experimental observations and plausible arguments will not stand up for long under expert questioning. If science is to evolve, it must continually purge itself of misconceptions (4.5), follies (3.6) and practical errors (3.5): there must be preserved a central store of absolutely reliable knowledge, from which to draw in evaluating novel ideas and on which, very slowly and carefully, to build. In order that science may retain its reliability and credibility, each scientist is expected to exercise critical vigilance over his own work and the claims of his contemporaries.

The psychological contradiction between these two normative principles is real, and cannot be reconciled by a bland formula. It is difficult for the layman to appreciate the in-built tensions of the scientific life – the conflict between *criticism* and *imagination*, between *normal* science and *revolutionary* science, between *preserving* the paradigm (4.4) and *breaking through it* (3.6). It is even more difficult to appreciate the social conventions and behavioural norms that keep such conflicts under control, so that they do not destroy communal cohesion.[13]

[13] The moment of truth for many young scientists comes when they first act as a referee for a scientific paper; having striven for years to get their own work published *against* the criticism of anonymous referees, they find themselves, by psychological role-reversal, on the other side of the fence. Thus do we eventually internalize the 'scientific attitude'.

The facts demonstrate, nevertheless, that a healthy scientific community can accommodate intellectual controversy without breaking down. Consider, for example, the history of Wegener's hypothesis of continental drift (4.5).[14] Here was an immense revolution of thought, grossly overdue despite the almost pathological conservatism which had to be overcome by its advocates. Yet by the standards of most human institutions, academic as well as overtly political,[15] it was a very gentlemanly affair, in which the evolution of scientific knowledge was not disgraced by silenced voices or broken heads. Although the majority of the leading geologists of the time were unconvinced by Wegener's theory, and no doubt cautioned their students against it, it was not suppressed and forbidden as a 'heresy'. There were several set-piece public debates and conferences on the subject, and books and papers supporting Wegener's interpretation continued to be published. It is likely that some of the protagonists did not fare quite so well in their academic careers as hindsight would now consider deserving; but when, eventually, new evidence from rock magnetism vindicated Wegener's bold and imaginative hypothesis, the 'old gang' were not swept into scholarly oblivion by the 'revolutionary' supporters of the new orthodoxy of plate tectonics.

This truly remarkable and civilized behaviour amongst scientists we take for granted (3.5): these are the standards against which occasional pathologies are judged. And if those who rule society – aristocrats or democrats, capitalists or socialists, conservatives or radicals – want scientific knowledge on which they can rely, they must not allow the inner tension of science to slacken, break, or overbalance. According to the narrow logic of bureaucratic planning, it is a wasteful, irrational system that ought to be made efficient and economical. But by encouraging innovation, yet conserving past achievement, by calling the gambling competitive spirit from each of us, yet making us also the guardians of truth and the judges of quality, it is remarkably successful as the source of many wonders.

6.4 *Keeping in touch with reality*

Unfortunately, the imaginative/critical tension of science is not sufficient, of itself, to justify belief. These features of the social model keep it growing and evolving, but do not determine the direction of that

---

[14] As outlined, for example by Hallam, *A Revolution in the Earth Sciences.*
[15] As counter-exemplified by *The Rise and Fall of T. D. Lysenko* by Zh. A. Medvedev (1969: New York: Columbia University Press), where the institutional framework of a scientific community was broken by state authority and trampled under the heel of a charlatan.

evolution. It ensures that scientific knowledge will be very closely fitted to the criteria by which it is selected and stored in the archive, but gives no guarantee as to the legitimacy of those criteria. If all scientists were convinced sectarians of a fundamentalist Church, then they would create a science in which Biblical authority would be a privileged argument. If they had been born and bred into Zande culture[16] where witchcraft is regarded as normal explanation of many daily occurrences, then they would see no reason to exclude such interpretations from their science. The model as it stands is thus open to all the objections of *cultural relativism* (5.10), which every scientist and most philosophers feel bound to rebut.[17]

Let us recall, however, that the 'scientists' in our model are supposed to act only as proxies for 'Everyman' (6.1). They would be betraying their trust if their selection criteria were not those of all mankind. Those who were not fundamentalists or born into Azande would be quite justified in rejecting a 'science' that was not fully consensual and would ask to be convinced by the evidence themselves before they felt bound to believe.

In practice, some degree of sectarianism can often be observed within science itself. The experts in a particular field can become so indoctrinated (6.2) and so committed to the current paradigm (4.4) that their critical and imaginative powers are inhibited, and they cannot 'see beyond their own noses'. In these circumstances scientific progress may come to a halt – knowledge may even regress – until intellectual intruders come through the interdisciplinary frontiers and look at the field without preconceptions. Thus it was only an influx of physicists drawing new evidence from the detailed study of rock magnetism that broke down the prejudices of the geologists against the theory of continental drift (4.5, 6.3, 6.7). In its extreme specialization, science is always in danger of fragmentation into small mutually incomprehensible and non-communicating domains, where standards of judgement and criteria of validity may fall into decay. Individuals who ignore the conventional barriers between disciplines and who are competent as experts in several adjacent fields, thus play a very important part in keeping those branches of science in touch with reality.[18] Sectarianism is a natural social tendency that cannot be

---

[16] E. E. Evans-Pritchard (1937) *Witchcraft, Oracles and Magic among the Azande* (Oxford: Clarendon Press).

[17] E. Gellner (1974) in *Legitimation of Belief* (Cambridge University Press) expresses powerfully the deficiencies of the 'sociology of knowledge' as a foundation for such objections, and arrives at essentially the viewpoint taken here.

[18] Polanyi (*Personal Knowledge*, p. 163) puts this point in a slightly different way 'nobody knows more than a tiny fragment of science well enough to judge its validity and value

entirely eliminated even in the world of objective knowledge, but it can claim no legitimacy, nor permanent institutional form, within the research community.

Nevertheless, as long as the practice of science as a whole is confined to a professional group who must undergo a long and rigorous training before they are recognized as qualified contributors or critics, there must always be the suspicion (6.6) that *they* are the ones who are out of touch, and that they are either all collectively deluded or are all in a conspiracy to keep the truth away from the people.

To rebut this charge, we must look again at scientific education (6.2). There is a significant difference between enlightened indoctrination and unscrupulous 'brainwashing'. The fundamental assumption of all science teaching is that the student has free, independent use of his own eyes, hands and brain, both to observe and experiment for himself and to grasp the theoretical interpretations of what he supposes he has seen. The evidence for the current scientific world picture is always presented *within the categorial framework of everyday realism* (5.10). He is not taught that science is an esoteric or specially privileged way of seeing things; he is told, in effect, that it is no more than '*commonsense writ large*'.[19] The trained scientist can thus act with some justification as a proxy for 'any reasonable man', because his paradigms of verification and proof remain essentially those of the everyday world.

When we think of the astonishing, counter-intuitive concepts of so much of modern science (4.4) – particle-like waves, wave-like particles, undying supercurrents, coded organisms, drifting continents – this must seem a strange conclusion. But if we study the actual path by which each such marvel was discovered and eventually incorporated in the scientific map, we see the reluctance with which even the beliefs of *naive* realism were successively shed. Although the modern scientist no longer believes (3.3) that 'when meat goes bad, maggots are engendered in it', he still uses the logic of sensorimotor coordination and natural language (5.8) to describe biological organisms and molecular structures, the geological events of bygone ages and the workings

at first hand. For the rest he has to rely on views accepted at second hand on the authority of a community of people accredited as scientists. But this accrediting depends in its turn on a complex organization. For each member of the community can judge at first hand only a small number of his fellow members, and yet each is accredited by all. What happens is that each recognizes as scientists a number of others by whom he is recognized as such in return, and these relations form chains which transmit these mutual recognitions at second hand through the whole community. This is how each member becomes directly or indirectly accredited by all.'

[19] K. R. Popper (1959) *The Logic of Scientific Discovery* (London: Hutchinson) p. 22.

of a clock. Generally speaking, mathematical formulations merely sharpen this logic (2.3), without superseding it. The world of science is not, after all, analogous to the world that Alice found when she went through the looking glass (2.9) but mostly follows the rules of identity and necessity that we all learnt as children.

We must not be misled by the mysteries of quantum physics and relativity, which seem so completely at variance with our experience (2.7). These conceptual schemes refer to phenomena that are seldom encountered by small children, and have not, therefore, had their inner logic incorporated in natural language (5.8). And it is an essential feature of these theories that they do not contradict everyday things. Quantum theory, for example, is taught as an *extension* of classical mechanics into realms that were previously inaccessible to observation, where the paradigms of naive realism – indivisible point particles and continuous wave fields – could not be assumed adequate. It is a matter of philosophical taste, not of scientific necessity, whether one emphasizes the 'correspondence' and 'complementarity' of the classical and quantum pictures, or whether one dogmatically asserts the primacy of the quantum theory 'map' (4.4) to describe macroscopic phenomena, where it gives precisely the same results as the classical theory.

From the history of science it is also quite clear that radical conceptual innovations are not accepted until all the orthodox interpretations have failed (4.5).[20] Max Planck, the founder of quantum theory, was gravely shocked by his own discovery – that energy was not infinitely divisible, but could be transferred only in discrete 'quanta'. Each scientific generation conceives itself as the guardian of the 'real world' mapped by its paradigms, and is reluctant to modify them except in the face of material facts and logical necessity. What looks like a frightening revolution to contemporaries may be seen in retrospect as an *evolution* in which the principles of rationality and realism continue to rule.

Science is thus to be distinguished from other systematic knowledge not only by the structure of its social institutions but also by its *metaphysic* or *ideology*. Various attempts have been made to define or characterize this ideology[21] but the whole of our present discussion suggests that this must be fruitless. All that can be achieved is a circular argument where the various actual categories of scientific knowledge are displayed in succession, in eventual justification of their own right to exist. As has been repeatedly emphasized, the constitutive

---

[20] This is the message of T. S. Kuhn, *The Structure of Scientific Revolutions*.
[21] See e.g. E. Gellner, *Legitimation of Belief* (1974: Cambridge University Press).

principles of 'scientific method' are not to be found by logical analysis of the messages in the scientific archive, but lie within the mental domains of the scientists and other people who give their collective assent to these messages.

The fundamental *continuity* and *coherence* between the sense of everyday reality acquired in childhood (5.10) and the reality of the scientific world picture is not an illusion. Science uses the same mechanisms as in the growing child[22] – sensorimotor coordination of observation and experiment, pattern recognition and the mental transformation of images, communication with a world of 'others' and tests to select consensual conceptual schemes (5.8) – not merely to implant this sense of reality into its practitioners but also as a means of acquiring uniquely faithful knowledge of the material domain. Despite numerous mistakes and misconceptions, the child constructs a picture of his immediate world that is fundamentally reliable and worthy of belief. It is difficult to deny the same qualities – and possible defects – to the picture of a much wider world created by the larger social instrument of science.

6.5   *How* much *can be believed?*

Science continually evolves (6.3). Scientific knowledge is under constant revision in the light of new evidence. From a practical point of view, it is not the ultimate truth of the scientific world picture that matters (5.6), but the scientific answers to particular questions, such as 'why is the sky blue?' or the degree or credibility of particular scientific theories, such as that biological ageing is due to accumulation of random mutations during successive cell divisions. Both layman and scientist are concerned about the epistemological *status* of the information in the scientific archives.

The concept of an *archive* of reliable scientific knowledge (1.4) is much too schematic.[23] There is no *Encyclopaedia* where *all* well-

[22] The forceful analogy between the intellectual strategies of research and the psychology of perception has been pointed out, for example, by N. R. Hanson, *Patterns of Discovery* by D. T. Campbell, *The Psychology of Egon Brunswick* and by R. L. Gregory (1973) *Proceedings of the Royal Institution*, **43** 117–39.

[23] 'The assumption seems to be that any well-equipped society catering to the enlightened interest of its members would make philosophical conclusions about knowledge publicly available, perhaps in reference libraries, or better, as an addition to the telephone service so that people could at any time or in any situation or stage of debate be given the authoritative view on the matter (we might sympathize with the unfortunate man entrusted with the task of compiling such conclusions. . . he would almost certainly recommend suspension of judgement in his letter of resignation. Strictly speaking, of course, any headway he made would only add to the confusion).' A. Naess, *Scepticism*, p. 111.

established science, and only well-established science, may be consulted. If such an institution existed, it would be in constant agitation, as new information was being added, and old facts and assertions struck out. And not even the Swedish Royal Academy, whose responsibilities include the award of Nobel Prizes, would be willing to take upon itself the invidious duty of deciding, from day to day, what should be in, and what should go out.

Amongst professional scientists, the corpus of what is called the *literature* of a subject consists of *papers* published in reputable *journals*, catalogued regularly in, say, an *abstract journal*. But the layman who attempts to consult all the papers relevant to a particular scientific question is soon wearied and appalled by the confusion and diversity of fact and opinion that he will find. At the research frontier, scientific knowledge is untested, unselected, contradictory and outwardly chaotic (6.3); only the expert can read, interpret and weigh such material.

In despair, the earnest enquirer might then turn to the lectures and text-books of the science teachers (6.2), where confidence will be restored by an air of clarity and certainty. Here, we suppose, are the deep foundations of knowledge, which are never likely to be disturbed. It is comforting to learn and accept the *constitutive principles*, the *categorial framework*, the *dominant paradigm* (4.4) of the current scientific world picture – even though history may suggest that these foundations are seldom quite so firm as teachers and students fondly assume (4.5).

But text-book learning is only a tiny fraction of what is quite well known and quite well understood. The results of recent research must surely have relevance to decision and action. Faced with a contemporary scientific question, such as 'Does mass fluoridation of water supplies have adverse effects on human health?' we seek the advice of an expert on the subject, or read our way into research monographs, review articles, colloquia and conference proceedings, not expecting a definitive answer but hoping to get an assessment of the 'consensual weight' of various points of view.

There again, in pursuit of certainty, the layman finds that science does not proceed by counting votes. Even a complete consensus is seldom publicly determined or proclaimed; the best we may expect is an answer that is said to be the 'almost unanimous opinion of the experts', backed by what they would describe as 'the overwhelming weight of the evidence'. Such an answer would certainly be reliable as a basis for immediate practical action, or as one of the links in a chain of reasoning concerning some other scientific problem. Yet observe the customary caution of the scholar, leaving room for the possibility of error, not precluding the need, perhaps, for further research on the subject. The scientist who is a little over-optimistic concerning the validity of his fundamental assumptions (6.2) is quite likely to hedge his bets when it comes to questions that he has himself helped to settle by his own research. He is all too familiar with deficiencies of technique and fallacies in the argument, and may give exaggerated weight to contrary criticism that he cannot easily rebut.

When there is open *controversy* between adequately matched scientific authorities, the spectator is advised to suspend judgement, sit back, and watch the fun. The drama of conflict rapidly draws other scientists into the fray, the subject becomes 'exciting' and fashionable – and the issue will soon resolve itself as a thousand papers bloom. On 'anomalous water', for example, the forces of orthodoxy were always strong enough to keep the issue open, and eventually to disconfirm the 'discovery' (3.6). No doubt the 'missing solar neutrinos' will soon be found, or else one or other of the competing theoretical explanations will come to the fore (2.9). It is seldom that such a question remains for a long period in the limbo of a Gödelian proposition (2.3) – neither provable nor disprovable by current research techniques. The controversy over continental drift (4.5, 6.3) was unusual in that it was not settled for half a century, despite serious attempts on both sides to arrive at a conclusion.

At the far end of the spectrum of credibility lies the speculative hypothesis, or unexpected observation that seems to contradict radically the current paradigm. When first reported – sometimes, nowadays, unprofessionally in the popular press – such a message stands on its own, with no more than the scientific competence of its author to recommend it. Until it has been supported or condemned by other reputable scientists, confirmed or disconfirmed by further experiments or become the football of an active controversy, it can scarcely be said to belong to the scientific domain. Yet such a singular message may be provocative, unsettling, a cause for serious thought. The layman may be no more at sea than the acknowledged expert, who may be called upon to pronounce on delicate questions of statistical inference, signal-to-noise ratios (3.5), unconscious experimental bias, implausible assumptions made in theoretical predictions, etc. The first announcement of a discovery (3.6) is always an exciting moment in science; yet until the provocative, unexpected message has been to some extent 'processed' by the social instrument of the scientific community it can neither be ignored nor taken as a basis for action, and thus lies outside the dimensions of scientific belief and doubt (5.5).

Although we must rely upon the 'scientific authorities', in their published works or in personal consultation, for information about the actual contents of the scientific archives, it is expecting too much of them to vouch for the truth of all that they tell us. What can be deceptive is the degree of credibility that the experts themselves assign to particular scientific propositions. Scientific knowledge is too vast in quantity, and is amassed and assessed too haphazardly, for anyone to have a sound perception of the relative status of all the little pieces of information that go to make a scientific discipline.

It often occurs, for example, that a conjectural *phenomenological* or *heuristic* model, not solidly based on other scientific data, is found to fit a particular set of experimental facts quite well (2.9). For such a theory to achieve full scientific validity, it should have shown its predictive power (2.8), or have been deduced by sound arguments from some better-established principles. But if these validating investigations have not been pursued, and if there is no competing interpretation of the observations in question, the model may be incorporated into the scientific map (4.2) as if it were an established truth. The unconscious assumption is that as the whole network of knowledge becomes more closely connected (2.10), as more knots are tied between loose ends, the original provisional interpretation will validate itself by the coherence of the whole theory. This can be a most dangerous phase in the development of a scientific subject, where inadequately tested hypotheses can establish themselves as details in a larger picture, and acquire much higher credibility than is justified by the research that has been actually done on them. It is very difficult for the expert, strongly influenced by the known reliability of the overall paradigm, to acknowledge this uncertainty in more specialized issues.

To appreciate the weakness of what may seem a well-woven tissue of scientific inference, one has only to look at the effect of a single false item of information. 'Piltdown Man', for example, was 'discovered' in 1911 and was always regarded as a mysterious and somewhat anomalous extinct species.[24] Yet the ancestral tree of humanity had

---

[24] 'Some experts still doubt whether a lower jaw which resembles that of a chimpanzee in several respects should be assigned to a skull which is purely human in its characters'. A. Keith (1929) 'Man, Evolution of' *Encyclopaedia Britannica*, 14th Edition.

to accommodate him[25] and was distorted accordingly by many anthropologists of the day. It was not until 1949 that a careful investigation exposed '*Eoanthropus*' as a crude hoax. The wording of the article by a leading expert in an authoritative volume is cautious enough in expressing considerable doubt about all its inferences, but does not perhaps make clear that the evidence then available in this branch of science was quite inadequate to validate any hypotheses beyond the level of 'wild surmise', and tries to suggest a theoretical coherence that could be quite misleading to the naive enquirer.

It seems unfair to look back over the history of science, and to focus on the occasional follies or errors of honest, industrious men and women. But how otherwise can we appreciate that under the serious, sincere prose there may be hidden pitfalls into which the scientist as well as the layman can easily fall. The scholar planning his research must estimate for himself the epistemological status of all that is relevant to his subject – the idiosyncracies of his apparatus, the significance of unexplained 'anomalies', the strength of the assumptions underlying theories, the boundaries of genuine ignorance, and the unseen difficulties that may stand in the way of further progress. The strategy of research is never blind trial and error, but is governed by the intelligent appreciation of possibilities and 'solubilities'.[26] To ask 'How *much* is to be believed?' is to invite oneself inside the scientific process, and to attempt oneself to contribute to the evolution of knowledge.

That is why the layman, anxiously seeking advice or justification from 'science', often comes away frustrated and sceptical. Why won't 'they' – the scientific authorities – tell him simply and clearly what he may or may not believe? Why do they fob him off with generalities, or vague possibilities, or contradictory speculations? This is not the place to follow scientific knowledge further from the noetic domain through the interface into the domain of *action*, where all the questions of application, of social responsibility, of science policy eventually originate.[27] It is important to appreciate that using the scientific in-

[25] 'The discovery at Piltdown shows that at the beginning of the Pleistocene period a race of man had come by a brain that had reached a human estate, and that this race still retained certain definite simian characteristics in its jaws, teeth and face... Man's history in Europe has been traced throughout the Pleistocene period by the discovery of his fossil remains and of his stone weapons... Primitive though these early Pleistocene Europeans undoubtedly were, we cannot withhold from them the right to be human... Such discoveries of fossil man as have been made lead us to infer that early Pleistocene times, so far as concerns our direct ancestry, was a period of rapid evolutionary change'. A. Keith, 'Man, Evolution of', *Encyclopaedia Britannica*.
[26] P. B. Medawar, *The Art of the Soluble*.
[27] See e.g. Ziman (1976) *The Force of Knowledge*.

strument is not so simple as consulting a dictionary for the meaning of a word, or solving an arithmetical problem on a slide rule. To discover what a scientist really knows, it is necessary to feel for oneself the critical/imaginative tension (6.3) that governs his mind and art.

6.6 Parascientism

Institutionalized, *collegiate* science (6.1) is always vulnerable to the charge that it is no more than an 'Establishment' that defends only its own orthodoxy. The consensus towards which science strives applies, in practice, only to the members of a scientific community already heavily indoctrinated in the current paradigms (4.4, 6.2). Despite all its high ideals and good intentions (6.3), such a community must inevitably resist radically new ideas that upset its hard-won position and throw into doubt all the earnest labours of its members (4.5). This charge is easily substantiated by reference to the familiar list of important scientific discoveries that were rejected by the established scientists of their day.[28]

To be fair, one should also remark that 'If you have had your attention directed to the novelties in thought in your own lifetime, you will have observed that almost all really new ideas have a certain aspect of foolishness when they are first produced.'[29] Most of the ideas that were initially rejected soon gained scientific credibility. History also records the successful defence of reliable knowledge against such 'discoveries' as Blondlot's 'N-Rays', the 'mitogenetic rays' of Gurwitsch, the 'Allison effect',[30] and, in more recent years, 'anomalous

[28] For example, Camille Flammarion, a French astronomer interested in psychical research, in his book *L'Inconnu et les Problèmes Psychiques (1900)* gives a typical list: 'Pythagoras conceived the rotation of the Earth about its axis, but met only scorn from the astronomers: Galileo was persecuted for having affirmed "the magnitude of the world system and the insignificance of our planet": Lavoisier had dared to show, against the opinion of his time, that the air was composed of two gases: the observation of Galvani, according to which a strip of tin and a strip of copper produced electricity, was greeted with an immense roar of laughter; in 1841, the Royal Society of Great Britain refused to publish the most important paper of Joule, and Thomas Young, founder with Fresnel of the wave theory of light, was covered with ridicule'. Flammarion finally recounts that at a meeting of the Académie des Sciences, on 11 March 1878 he saw the physicist Du Moncel present Edison's phonograph. 'Once the presentation was finished, the apparatus was set to docilely reciting the sentence recorded on the cylinder. Then I saw an academician of ripe age, his mind filled, indeed saturated, with the traditions of the classical culture, nobly indignant at the audacity of the inventor, throw himself on the representative of Edison, and seize him by the throat crying "Scoundrel! We will not be the dupes of a ventriloquist".' M. Blanc in *La Recherche*, **7**, 387 (1976) (my translation).
[29] A. N. Whitehead (1925) *Science and the Modern World* (New York: Macmillan) p. 49.
[30] I. Langmuir (1953) *Pathological Science*: unpublished tape recording, transcribed and privately circulated by R. N. Hall.

water' (3.6). The scientific literature is a mine of similar reports that are not confirmed. It is implicit in the very nature of science that there should be such episodes, where the delicate balance between imagination and criticism (6.3) becomes evident. There is no substantial evidence that scientific knowledge would evolve more rapidly, or would be significantly more reliable, if there were a radical change in the point of equilibrium between these two forces.

Nor is it obvious that the contents of the scientific archives would be improved if all new knowledge had to be approved by, so to speak, *juries* of lay persons previously unacquainted with the complexities of the issues involved. It is good that science journalists and independently minded authors, such as Mr Arthur Koestler, bring to public attention some of the controversial questions of science, but it is not clear that public debate resolves the underlying intellectual issues. The serious critic inevitably begins to acquire some of the expert knowledge of an 'insider', and becomes an adjoint member of the scientific community rather than a genuinely independent assessor. The whole problem of public participation in technical policy-making is of the greatest political urgency, but it suggests that a much wider diffusion of 'what the scientist knows' will be required before it can be reflected back into an improvement in the quality of that knowledge.[31]

Surrounding the relatively well-defined boundaries of 'official' institutionalized science, there naturally develops a fringe of *parascientific* ideas seeking formal recognition by the scientific community or by the public at large. Since the Republic of Science must never be closed (6.4), it is important that public justice be done to such ideas in open court.[32] Nothing is more destructive to the credibility of science in the public eye than the appearance of scorning or suppressing anything that might in any way be considered a sincere, and possibly valuable, contribution.

It is often argued that the best policy would be to open the communication system of science to the full, trusting that, in the end, the truth will triumph. But this would not only be very damaging to the internal critical apparatus of conventional science, whose function is absolutely essential (6.3) in maintaining a reliable consensus: it would also fail to satisfy the exponents of unorthodox views.

The difficulty is that the parascientist usually asks for more accre-

[31] Leaving aside the key issue of 'science policy': what research should be *initiated*, towards the resolution of what questions?
[32] For this reason, it was quite wrong for certain astronomers to put pressure on Macmillan's for publishing the work of Immanuel Velikovsky: see *The Velikovsky Affair* edited by A. de Grazia (1966: London: Sidgwick & Jackson).

ditation than just to be 'heard'. The complicated biography of Galileo has been simplified into an epic of 'science against prejudice'; the would-be hero is not satisfied with publication of his theory in a reputable journal but also insists on an advance on the posthumous acclaim that he feels sure he deserves.[33] And it must be admitted that there is a certain attraction in the sight of a lightly-armed David attacking the Philistine ranks of academia. It is very likely that the 'outsider' gets most of the sympathy, and gains far more credence than his cause usually deserves.[34]

A very typical case is the professional scientist of some standing who goes gently round the bend, and begins to expound some incoherent irrational theory.[35] Beyond a certain point it soon becomes clear to his colleagues that nothing is to be gained by arguing with him. But it is very difficult to prove that he is insane. When a bank manager goes dotty, and hands out five pound notes in the street, claiming to be a reincarnation of the late Mr Paul Getty, there is not much doubt that he is unfit for his job; but a famous professor, respected for his brilliant insights, long experienced in the byeways of his subject, can conjure up a fascinating scaffolding of special pleading and plausible conjecture to support his obsession. The only professional colleagues competent to diagnose his condition can be accused of prejudice or jealousy. The scientific community cannot altogether deny him the right to publish in the recognized journals, although it is usual to adopt a tacit policy of avoiding critical comment, to keep the peace. Unfortunately, this often exacerbates the paranoic symptoms, so that the scientist may, so to speak, exile himself from his former community, issuing letters to newspapers and publishing books in which, say, the President and Council of the Royal Society are accused of betraying their sacred trust by refusing to endorse publicly his profound insight.

[33] Whereas, in truth 'A man cannot wear the mantle of Galileo simply because he stands against an establishment that treats him badly; he must also be right, or at least brilliant. If he isn't, his story will probably become the farce that Marx recognized as the historical repetition of tragedy – Galileo the tragedy, Velikovsky the farce.' S. J. Gould (1972) *Science,* **176**, 623 – reviewing *The Case of the Midwife Toad* by Arthur Koestler.

[34] 'There are spirits in our subconscious that make us believe. The spirit of paranoia prods "Believe it because all those scientists are picking on it". The spirit of hypocrisy urges "Pretend to believe it because it can gain you attention and wealth". The spirit of conservation says "Believe it because it fits with other things you believe" (or more often, "don't believe it because it doesn't fit with other things you believe"). The evangelical spirit whispers "Believe it because you want something to defend".' W. K. Hartmann (1972) in *UFO's: A Scientific Debate* edited by C. Sagan & T. Page (Ithaca, New York: Cornell University Press) p. 20.

[35] In this matter, I forbear to cite chapter and verse!

Such episodes, although privately distressing to friends and colleagues, and tiresome to editors, referees, book reviewers and others with responsibilities in the communication system of science, do not, perhaps, seem of great importance. Yet the manner in which they are handled may have considerable influence on the public 'image' of science, and under pathological political conditions may be highly relevant to our theme (6.3). What happens to scientific knowledge, we may ask, in a totalitarian regime, when one of the most senior scientists in the country falls under such a delusion?[36] Who will guard this guardian?

On the other hand the isolated 'crank' without scientific training is not likely to cause much damage to the scientific community because he has no official platform from which to propound his views. But for this very reason, it is proper that his claims be examined seriously by competent experts, in case there is something in them. Since there is no official accrediting agency for 'scientific' knowledge (6.5) this responsibility falls on the editors and referees of reputable scientific journals, who are often put to considerable trouble pondering over some scarcely comprehensible communication before advising that it must be rejected. This is not to say that the author is insane. The fundamental difficulty is that the lay person with a bright idea usually has insufficient command of the standard consensible nomenclature (Chapter 2) of the subject, and is likely, therefore, to express himself so obscurely that the expert cannot easily decide whether it is naively unsound or cryptically deep.[37]

It is also characteristic of such material that it attempts to answer some paradoxical question, such as the relationship between space and time, or the complementarity of wave and particle, that has become publicly notorious as a controversial scientific issue. The untaught author has no conception of the depth and complexity of the scientific discussion that by now surrounds such an issue (2.10, 4.2) but sees himself as another Einstein[38] taking a bold, brilliant leap into the unknown.

[36] This is not a hypothetical case. I have observed it personally in an advanced country with a thriving scientific community. Fortunately, the damage was kept localized by a skilful conspiracy of more junior scientists.

[37] J. J. Waterston's paper (1845) adumbrating the kinetic theory of gases was rejected by a referee as 'nothing but nonsense', and J. Mayer's work (1842) on the principle of the conservation of energy was initially refused publication in a physics journal. Waterston was an engineer, Mayer a physician, and neither was a master of the conventional language of mathematical physics.

[38] 'In the future will not Zweistein devise another pattern, followed later by Dreistein's, and so *ad infinitum*?' wittily asks W. H. George (1938) in *The Scientist in Action* (London: Scientific Book Club).

The fact is, of course, that almost all such earnest endeavour, whether in the form of a handwritten letter, or as a handsomely printed volume under the imprint of a major publisher, is quite worthless. The writings of Immanuel Velikovsky for example, despite their enormous sale and the popular cult that has grown up around him, offer no serious challenge to science – for they do not accept the primary principles of consensible evidence and consensual interpretation on which the whole scientific enterprise depends.[39] Belief in such theories must rest, therefore, on other foundations than those that underpin scientific knowledge. Yet the professional scientist, in all the panoply of his learning, must remain alert to the possibility that the strange communication from an official of the Swiss Patent Office comes from another A. Einstein, and that the minor clerk of the Madras Port Trust who introduces himself so humbly is S. Ramanujan.[40]

It is important for science that neither good ideas nor significant *phenomena* should be ignored. Advocates of research in *parapsychology* argue that the various aspects of 'extra-sensory perception' (ESP) have been seriously neglected by scientific observers.[41] These observations are admittedly equivocal, unpredictably fragmentary, and to many scientists totally unconvincing; but on the face of it there seems no reason why they should not be brought within the frontiers of official science[42] and supported in the same way as other fields of research.

Nevertheless, there are grounds for caution in following such a development. Serious research in parapsychology is almost always motivated by more than the 'disinterested search for truth'. Those who undertake it[43] often make reference to 'psychic phenomena' as if guessing the symbols on packs of cards at a rate slightly higher than allowed by chance were necessarily connected with anecdotes of messages concerning the death of distant relatives, premonitions of disaster, or other essentially different and scientifically unmanageable events. The statistical data are interpreted according to models of 'telepathy' and 'precognition' which are not very different from those

[39] See e.g. R. Gillette (1974) *Science*, **183**, 1059–62; M. W. Friedlander (1975) in *Boston Studies in the Philosophy of Science*, **32**.
[40] Whose astonishing mathematical genius is recalled by e.g. J. R. Newman (1956) in *The World of Mathematics* (London: Allen & Unwin) pp. 366–76.
[41] See e.g. J. L. Randall (1975) *Parapsychology and the Nature of Life* (London: Souvenir Press) for a sympathetic account of this subject.
[42] In 1938, 89% of the members of the American Psychological Association thought that ESP was a legitimate subject of enquiry, though only 8.8% thought it was a 'likely possibility' or an 'established fact'. By 1952, this latter proportion had almost doubled to 16.6%. Randall, *Parapsychology and the Nature of Life*.
[43] E.g. J. B. Rhine (1937) *New Frontiers of the Mind* (1950: London: Penguin).

of primitive magic. And it is characteristic of much of the literature on the subject that credence is given to claims of psychic powers – for example, those made by Mr Uri Geller[44] – that could equally well be those of an expert conjurer.[45] In other words special attention is being given to observations that are defiant of the elementary principles of everyday realism (5.10) – seeing through solid walls, reading other people's thoughts, etc., without the aid of any special apparatus. It is as if there were a desire to discredit commonsense, to say that the world revealed to the senses is all right in its way, but sometimes it just ain't so. From a purely personal point of view, there may be nothing against such a philosophy of life; but it is not compatible with science as a body of *public* knowledge, whose 'objectivity' (5.6) and 'reality' (6.4) depend on the acceptance of everyday realism and its characteristic logic. We must not be deceived by the astonishing phenomena disclosed under the microscope of particle physics (4.4, 5.10): the simultaneous creation of particle and antiparticle out of radiant energy is certainly very damaging to *naive* realism, but it neither justifies, nor is consistent with the belief that metal spoons can be bent by the 'power of the mind', without other material influences. The recognition of psychic phenomena in our own individual lives is part of the human condition, but it is neither consensible nor consensual, and to confound it with science is to put both personal and public understanding in jeopardy.

Nor is it the bounden duty of 'science', as an institution, to take cognizance of, and to provide an explanation for everything that is thought to have happened in the world. 'Anomalous' events are often important as clues to discoveries (3.6); but in default of a theoretical model of all things, they are potentially infinite in number. It is the inspiration of genius to select for careful investigation just those apparently inexplicable phenomena that will provide material for new, consensual understanding. There are extraordinary things to be seen in the heavens, and those who watch them may be greatly enlightened by the aid of good science.[46] But the failure of a particular class of scientific concepts to explain a particular set of observations[47] does not provide sufficient grounds for 'falsifying' almost all our scientific

---

[44] E.g. R. Targ and H. Puthoff (1974) *Nature*, **251**, 602–7; J. B. Hasted, D. J. Bohm, E. W. Bastin, and B. O'Regan (1975) *Nature*, **254**, 470–2; J. G. Taylor (1975) *Nature*, **254**, 472–3.

[45] J. Hanlon, *New Scientist*, 17 October 1974, 170–85.

[46] For example, in the marvellous book of M. Minnaert, *The Nature of Light and Colour in the Open Air* (1954 edition: New York: Dover).

[47] Such as UFOs. Sagan and Page (eds) *UFO's: A Scientific Debate*.

understanding of a great many other phenomena, and it certainly does not make convincing a wildly speculative hypothesis introducing spaceships from extra-terrestrial civilizations and similar flights of the imagination.

Despite its demand for 'recognition' by a science whose mantle of intellectual authority it would like to assume for itself, *parascientism* has roots in the human psyche, and a role in human culture and social organisation, closer to those of revealed religion than of natural philosophy (1.1). Not only does it postulate spiritual forces and alien beings of immeasurable power, analogous to the hidden benevolences and malevolences of the ancient gods: it also ministers to the all too human desire to hitch oneself to the mysterious factors in life that seem beyond the control of the harsh necessities of commonsense and rational knowledge. What magic does for Azande (6.4), giving meaning to inexplicable ill fortune and disease, so do ESP, astrology and other cults for modern civilization. This service is by no means necessarily evil or deceptive amidst the vicissitudes of life.

But parascientism is a dangerous disorder for the experienced scientist, tending to lower his sceptical guard, and often bringing out an extraordinary capacity for credulity and self-deception. To those who are afflicted by it, the only answer of the scientific community must be: come with reliable consensible evidence, come with sound argument, and we will be ready to be convinced – but until that day you must not expect us to put much faith in your claims, nor to give much support to a cause in which we do not really believe.

6.7   *The limits of thought*

By the standards of a future epoch, the evolving science of today will seem gravely imperfect. But since that future is unknown, there is no means of assessing our present errors and ignorance. It is not difficult, of course, to point to whole areas of natural phenomena which we do not at present understand, but which we believe that a future science may penetrate. The most sincere account that we can give of the attempt to build a science of human behaviour (Chapter 7) emphasizes ignorance rather than reliable knowledge. More specifically, however, to make a rational assessment of our ignorance on a particular topic – to identify enigmas and formulate consensible questions – is itself an important scientific activity, scarcely to be distinguished from the phases of conjecture and unexpected discovery that are the beginnings of knowledge. Thus, for example, had the question been asked seriously 'What might be the appearance of the landscape of Mars?' the answers would surely have anticipated the astonishing observation from a space-probe that it was, like the Moon, covered with meteoric impact craters.

Any discussion of the present limitations of our knowledge is thus deeply rooted in present perceptions, and very rapidly enters the technical domain of the science in question. Nevertheless, we may legitimately speculate on the limitations imposed on science itself by its own structure, as the product of a community of human beings.

The fact that scientific knowledge belongs in the noetic domain (5.5), in the form of a message (2.1) or a map (4.1), imposes a severe restriction on the *quantity* of information that it could actually contain. No more than an infinitesimal fraction of all world events, of all potential data, can be recorded in the archive (4.2, 6.5). The more perfect science of the future could never approximate to the mind of Laplace's God-mathematician who knows the coordinates of every particle in the Universe and hence computes their future behaviour.

One is impressed, of course, by the enormous capacity of the modern electronic computer to crunch numbers at an immense rate, and hence to simulate the behaviour of quite complex real systems. Consider, for example, the technical feats of computational meteorology, by which the weather is predicted over a whole continent for a few days. But the input data, recorded at wide geographical intervals, although numbered in millions of 'bits', are only a tiny sample of what is truly relevant to what may eventually happen at the Sunday school picnic at Weston-Super-Mare on 27 June 1978. In any case, neither these data, nor the electronic signals into which they are transformed within the computer, nor the output weather map, are of any particular scientific interest in themselves.

Yet the fantastic power of the computer *at its own specialized tasks* (5.2) quite exceeds the capabilities of the unaided human brain. This poses a subtle challenge to science. As we here define it, scientific knowledge exists not only in the public archive as a recorded message but must have its counterparts in the minds of the scientists who know it. It is clearly unrealistic to assert that the meteorologist 'knows' all that is going on within the circuits of his computer; the quantity and complexity of the logical transformational processes that take place in the computation could not be reproduced mentally in a thousand years. Yet we say that we 'understand' the weather, not because by looking over today's meteorological chart we can *cerebrate* tomorrow's, but because we do indeed achieve such a prediction through the instrumentality of a material system, whose workings and programmes we only grasp piecemeal.

It is in the nature of mathematics (2.4) to generate logical consequences from given conditions by a long succession of steps that could not be intuitively grasped from beginning to end as a single 'thought'. Generally speaking, however, the mathematician learns to articulate the detailed elements of a proof into an overall system whose behaviour can be seen to conform to more general laws. Thus we may legitimately affirm that we 'understand' the weather in terms of quasi-invariant patterns such as 'depressions', 'anticyclones' and 'fronts', whose movements may be approximately predicted on rational principles.[48] This is the rationale of the familiar hierarchy of the categories of nature – the *nucleus* made of nucleons, the *atom* of a nucleus and electrons, the *molecule* of atoms, the *organelle* of molecules, the *cell* of organelles, the *organism* of cells and so on. Whatever one's philosophical attitude towards *reductionism*, there is the inescapable scientific necessity of trying to 'understand' and 'explain' the behaviour of any system in terms of a relatively few comprehensible elements without recourse to an elaborate extra-cerebral computation (5.4).

But the challenge of the computer resides in the fact that it is often successful in predicting the behaviour of genuinely complex systems without recourse to the 'mental' device of abstracting quasi-invariant features or factors. Experience has shown that quite simple sets of mathematical equations may have extraordinarily complicated solutions whose existence was previously quite unsuspected, and whose behaviour cannot be grasped intuitively.[49] On the other hand, the logical steps required to demonstrate some apparently simple proposition, such as the famous *four-colour theorem*, may turn out to be so numerous and complicated as to defy any other 'comprehension' than that provided by many hours of the inner workings of a large computer. We thus approach a strange frontier of science, where instruments seem to be taking over from human beings, not only in the collection of observations (3.4) but also in the processing and transformation of the scientific messages to give them meaning.

The key factor in the evolution of science remains, nevertheless, the power of the human mind to recognize a comprehensible pattern in a mass of detail (3.2, 5.1). This facility has its limitations; we may

[48] In fact, the discovery of the phenomenology of such patterns preceded success in solving the equations; the computational programme did not work until it was written in terms of a mathematical variable – 'vorticity' – which is typical of the observable cyclonic singularities of the atmospheric motion.

[49] R. M. May (1976) *Nature*, **261**, 459.

reasonably ask whether scientific knowledge itself is fundamentally restricted by these limitations. The data acquired by industrious instrumental observation (3.4) accumulate uselessly in the archives until they can be *mapped* scientifically (4.2). But making a map implies selection and arrangement of the survey data according to some organizing principle – a *framework* of *coordinates*, a *conceptual scheme*, a *paradigm*, a *theory*, a *hypothesis*. Such a scheme may, initially be purely private, internal to the mental domain of the mapmaker (5.4). But the consensuality of science insists that the patterns perceived by the innovator must be made visible to other scientists, and eventually acquire reality (5.10, 6.4) in the public noetic domain.

The *history* of science revolves always about the same question: *what are the origins of conceptual novelty?* We have gone out of our way to emphasize the social obstacles to revolutionary changes of outlook, and the resistance of the scientific community to innovation (4.5, 6.2). Yet the ethos of science is receptive in principle to such upheavals, and successful innovation is cherished and honoured (6.3). To conclude this chapter on a more positive note – and to come back to where science begins, in the minds of men – we might consider some of the mental processes that seem to act in at least the few rare individuals where 'conceptual mutations' arise.

To exemplify these processes, we return once more to visual perception (4.3, 6.4), as a metaphor or analogue of cognition in general. How is it that we do not at once 'see' what ought to be 'seen' in the information available to us? Why does it require an extraordinary individual talent to recognize, for the first time, a pattern that can be made obvious to the dullest student once it has been pointed out.

Habitual association of features inhibits the imagination.[50] The current scientific map or picture (4.4) bears with it conventional associations that are difficult to break (Fig. 25). Thus, for example, the brilliant hypothesis of Yang and Lee (3.3, 3.6) had to break through the 'obvious' proposition that 'parity is always conserved in elementary particle interactions' which had become so closely linked with the mathematical operations of quantum field theory as never to have been questioned before.

It may be, indeed, that the current paradigm completely misinter-

---

[50] 'Even if we drew one of these Indians with white chalk, he would seem black, for there would be his flat nose and stiff curly locks and prominent jaw...to make the picture black for all who can use their eyes. And for this reason I should say that those who look at works of painting and drawing must have the imitative faculty and that no one could understand the painted horse or bull unless he knew what such creatures are like': quoted from Philostratus'; Life of Appollonius of Tyana; E. H. Gombrich, *Art and Illusion*, p. 182.

Fig. 25. The mask on the right is viewed from the interior – yet can only be seen as a normal face. Picture by courtesy of R. L. Gregory.

prets a wide variety of facts, yet the truth cannot be immediately seen. Come to this picture (Fig. 26) with the well-learned preconception that it represents an *old* woman, and the longer you look at it, the less easily can you see it quite differently, as a young woman.[51] Elements that have a clear and obvious interpretation in the first case – for example, the lips of the old woman – do not disappear but acquire an entirely different significance in the new view of things. To make a *paradigm switch* on the scale of the revolution of the Earth Sciences in the 1960s (4.5, 6.3, 6.4) one must *unlearn* much that one thought one knew, to the extent almost of returning to the ignorance of the student or child.[52]

The case of continental drift is unusual in that a great deal of evidence had already been amassed that could immediately be reconstrued according to the new scheme. The power of *context* is illustrated by the famous trick question: 'which figure is the taller?' in Fig. 27. The 'correct' answer is 'they are both the same height' – as may

[51] The same example was used by N. R. Hanson, *Patterns of Discovery*.
[52] 'If we are to believe Wegener's hypothesis we must forget everything which has been learned in the last 70 years and start all over again' said a critic of continental drift (Hallam, A., *Revolution in the Earth Sciences*, p. 113).

Fig. 26. 'My wife and my mother-in-law.'

Fig. 27. From Day, R. H. (1972). Visual spatial illusions: a general explanation. *Science*, **175**, 1335. © Copyright 1972 by the American Association for the Advancement of Science.

Fig. 28.

readily be determined with a ruler. We are thus invited to be astonished at the deception that can be practised on the eye by a misleading context.

But hold on a moment! Considered as 'figures' – i.e. as representations of human beings – these are certainly not the same 'height'. The identical blobs of ink only have that meaning in the context of the rest of the picture. The imaginative step here is not to make measurements, to make allowances for an optical illusion, etc., but to pose the question in an alternative form, in a different universe of signification. Was not this what Einstein did, at the age of sixteen[53] when he imagined himself pursuing a beam of light with such a high velocity that it would seem like a spatially oscillatory electromagnetic field at rest – and thus came to realize the necessity for a principle of relativity which would forbid such an unphysical phenomenon?

If context can mislead, then absence of context can be deeply puzzling. What is the meaning of Fig. 28? I have shown this picture many times to a variety of audiences, and get diverse answers – a tree silhouetted against a forest fire, a stroke of lightning, blood vessels,

[53] See G. Holton (1973) *Thematic Origins of Scientific Thought* (Cambridge, Massachusetts: Harvard University Press) p. 292.

nerve processes, a river delta, etc. It is only when you have consulted footnote 54 below that you will be able to 'see' it for what it is. At its most enigmatic, science accumulates observations that are so out of context that they seem to have no meaning. When the radio astronomers began to observe 'quasi-stellar objects' (*quasars*) they could not decide whether these were truly at very great distances, as suggested by the large red shift in their spectra, and hence were of stupendous energy, or whether they lay at galactic distances with some unusual mechanism to explain the red shift. Until this uncertainty could be resolved, the whole concept of a quasar lay open, and beyond intellectual assimilation.

Nevertheless, the suggested interpretations of Fig. 28 are by no means fanciful. To say that this pattern resembles a *tree*, where each *branch* divides into further branches that seldom meet or cross one another, is consensible for almost anyone who has ever seen a normal tree and understood the way it grows. This sort of pattern is found in many different fields of science – not only in the flow patterns of rivers or of blood vessels, but also in the successive processes by which neutrons multiply in a nuclear reactor, or in the development of 'showers' of cosmic rays, or in the representation of kinship relations, or in the origin of species by organic evolution. It is scarcely necessary to make a formal analysis in the mathematical language of *graph theory*, where a 'Cayley tree' is defined as 'a graph without any closed circuits', to recognize and exploit a fundamental equivalence between certain theoretical schemes in a diversity of scientific disciplines.

Deeper than *models, metaphors* or *analogies* (2.5) lie the 'fundamental presuppositions, notions, terms, methodological judgements and decisions...which are themselves neither directly evolved from, nor resolvable into, objective observation on the one hand, or logical, mathematical and other formal analytical ratiocination on the other hand'.[55] Of all the intellectual resources of humanity, such *themata* are at once the most powerful – and the most restricting. The fundamental questions and explanations of science depend upon our interest in, and grasp of, what is expressed by words such as *atom* and *void, force* and *structure, symmetry* and *conservation*. These are the materials – personal and intuitive on the one hand, intersubjective and consensual on the other – out of which theories, interpretations, meanings and realities are constructed. Whatever may constitute the external world,

[54] Fig. 28 is an infra-red aerial photo of a river delta, reproduced by permission of the Aero Service Division, Western Geophysical Co. of America.
[55] Holton, *Thematic Origins of Scientific Thought*, p. 57.

whatever the sense impressions that crowd into our heads, our scientific maps and pictures can be made only of elements selected and filtered through the thematic net.

This Kantian critique of scientific knowledge is a consequence of the way in which, as children, we acquire our basic consciousness of living and being (5.8). It is easy to recognize in some of our scientific themata the elementary categories of sensorimotor action and natural language. An 'atom' is an object that persists through time, like the toything that only temporarily 'disappears' in one of Piaget's experiments with small children. Symmetry principles have their origins in the beauty of a flower, or the tesselations of a tiled floor. Modern particle physics is delightfully anthropomorphic: look at the 'life story' of the 'strange particle' in a bubble-chamber photograph such as Fig. 11.[56] To understand the 'annihilation' and 'creation' of particles it is as necessary to have read fairy tales and murder mysteries as text-books of analytical mechanics.

The thematic analysis of scientific thought has scarcely begun. It could be, as Holton sometimes suggests, that we only have at our disposal a finite set of themes, as old, perhaps, as the human mind. If that were indeed the case, then scientific knowledge would be correspondingly restricted; there might be some aspects of the material domain which, like colour-blind men, we could never become aware of, never communicate to one another, never hypothesize, theorize, model, falsify, or confirm.

The history of science, however, gives grounds for cautious optimism on this point. Darwin, in his theory of evolution, drew upon a thema – Malthus's principle of the expansion of population and the struggle for survival – which seems to have been an entirely novel product of the social conditions of the day.[57] We, in our turn, use Darwin's theory of evolution through natural selection as a thematic principle in a wide variety of contexts in the natural and social sciences. It is difficult to find historical equivalents or precursors of other such fundamental principles as Einstein's conception of the force of gravity as a geometrical constraint, or the 'Copenhagen' interpretation of quantum mechanics in terms of probability amplitudes.

At every level, from the grand scale of elementary-particle theory to the minutiae of biochemical reactions, scientific thought is a com-

[56] Holton, p. 106.

[57] That is, if we are to believe R. M. Young (1971) in his contribution to *The Social Impact of Modern Biology* (London: Routledge & Kegan Paul). He would like to taint biology with the tarbrush of reactionary politics: I see this as a typical example of the scientist as *bricoleur*, picking up any suggestive theme as possible material for a theory.

pound of old themes and new concepts, major and minor, fantastical or pedestrian, which combine to transcend the thought of the past. Science itself continually generates new themata that appear as organizing principles in the new world pictures (4.4), the new aspects of reality (6.4), that scientists create and share. Since we cannot possibly foresee the patterns of future human thought, we cannot rationally circumscribe their potentialities in the noetic domain.

It is sobering to note, nevertheless, that many of the most imaginative themes of Western Science – for example the so-called *bootstrap principle*, by which all 'elementary' particles are conceived as 'elements' of one another – are close in spirit to the ancient intuitions of Eastern mystical philosophy.[58] Faced with observations in which he can discern no pattern, the scientist of the future may be wise to fall into meditation, or perhaps take psychotropic drugs to acquire an 'altered state of consciousness',[59] wherein some of the aspects of reality to which his primary senses and the deeper levels of his brain are attuned may float free of the inhibitory nets of verbalization to reveal new ways of scientific thought which will not be so personal as he first imagines but which may find their echoes in the minds of other men and women.

[58] Capra, *The Tao of Physics*.
[59] See C. T. Tart (1972) 'States of Consciousness and State-Specific Sciences' *Science*, **176**, 1203.

7
Social knowledge

'when men, in treating of things which cannot be circumscribed
by precise definitions...talk of power, happiness, misery, pain,
pleasure motives, objects of desire, as they talk of lines and
numbers, there is no end to the contradictions and absurdities
into which they can fall'.

Macaulay

7.1 A science of behaviour?

The question is: can we acquire reliable, consensual knowledge about
human behaviour? Since mankind exists to the full only in the company
of his fellows, this is a demand for a *science of society*. Do we have such
a science? If so, how much of it is to be believed?

Many natural scientists deny scientific status to intellectual disciplines such as *social psychology*, *sociology*, *anthropology* and *economics*. But
there is nothing in our basic model of science (1.4) to say that the
behaviour of human beings cannot in principle be studied by the same
methods as the behaviour of neutrinos, nucleic acids, or nematodes.
A great many highly intelligent and well-informed people do, in fact,
devote themselves to research in what *they* call the *social* or *behavioural
sciences*. In the face of the enormous amounts of mental and material
efforts that are thus spent, it is important to assess these activities and
the body of knowledge they produce, without initial prejudice.

The claim to scientific status for these disciplines is not simply a form
of words; it is intended to convey correspondingly high credibility. The
general conclusion that we arrived at in the last chapter is not that
science is 'absolutely true', but that within its limits scientific knowledge
is as reliable a guide to action as is ever to be found. Within his own
field, in relation to questions to which scientific answers have been
given, the scientist has a legitimate claim to expert authority which
cannot easily be disputed by the layman (6.5). Thus, the question
whether a particular body of knowledge or academic discipline is
genuinely 'scientific' is severely practical; upon its answer may depend
our lives, our fortunes, our sanity, or our happiness. It is on this very
issue that the epistemological challenge (1.1) is at its most serious and
crucial for any responsible person.

Discussion of this issue easily gets bogged down in a mass of detail.
We cannot possibly assess every item in the vast archives amassed by
the social scientists in the course of their research. Nor would the

question be in doubt if economics were merely the compilation of statistics of prices, profits, purchases and payrolls, or if anthropologists confined their activities to the collection of cultural artefacts and kinship categories. The consensibility and potential consensuality of material facts associated with human behaviour can be taken for granted in the same spirit as we ascribe scientific status to material observations (Chapter 3) relevant to the natural sciences.

To believe in science is to have some confidence in its *predictive* power (2.8, 5.5). To make a reliable prediction, it is necessary to have in mind a sound *model* (2.5) or *map* (4.2) or *picture* (4.3) of that aspect of things. The credibility of any science of behaviour thus depends on the status of its *theories* – the selection and communication of the observational data, their mental organization into significant patterns and the validation of conjectures and hypotheses by the collective activity of the scientific community. The fundamental issue for the behavioural sciences is whether the theory-building process can produce a strong, sure, unambiguous framework of concepts and relations as reliable in its own domain as the physical and biological sciences in theirs.

It must be emphasized, however, that man does not need 'science' in order to live. From time immemorial, tradition, emotion, poetry and myth have provided him with comprehensive schemata of belief and motive. Until quite recently, humanity has managed to survive quite well, thank you very much, without the benefit of any consciously scientific study of its own behaviour. What we ask of a science of society is a body of knowledge, a guide to action, that is significantly more reliable, significantly broader and deeper in scope, than the agglomerations of practical wisdom with which most of what we do is still decided. The concern of this chapter must be, therefore, with the status of fairly general theoretical systems or research methods, whether now in existence or potentially to be discovered, that might illuminate the deep and paradoxical realities of human social life. The truisms of *descriptive* sociology can speak for themselves.

7.2 *Categorial imprecision*

Over the centuries, the physical and biological sciences have developed effective techniques of consensible observation (3.1), unambiguous media of communication (2.2), high standards of criticism and validation of theories (2.9) based on a strong metaphysic of everyday realism (5.8, 6.4). It is natural to subject social and psychological phenomena to these same methods of enquiry.

The traditional aspiration of sociology is that it should become a science like *physics* – that it should one day 'find its Newton'. But physics is a highly specialized discipline, given over to the mapping of the material domain in the idealized consensual language of mathematics (2.7). If we are ever to construct a genuine 'psycho-physics' or 'socio-physics' we must first find the means to represent psychic entities or social behaviour in a precise logico-mathematical language. This poses fundamental problems of two kinds.

Even in the mathematical representation of the *material* domain, *unsharp* categories obeying *three-valued* logic have to be replaced theoretically by *sharp* categories obeying the much simpler *two-valued* logic of ordinary mathematics (2.6). In the physical sciences this approximation is permissible because it is applied only to carefully chosen subject matter under highly contrived conditions. The errors in the outcome of a mathematical analysis are thus tolerable and do not vitiate comparison with experiment or observation (2.9). But the gap between the logic of experience and the logic of theory cannot be bridged by wishful thinking; the characteristic scientific activity of *theorizing*, of *model-building* (2.5), of *mapping* out our knowledge (4.2), is founded upon the recognition in nature of reasonably sharp categories to which deductive argument can be applied with reasonable certainty.[1] Without such categories, conventional scientific method is hamstrung.

Indeed, in all branches of the natural sciences, whether or not they are amenable to mathematical analysis, the problem of *classification* is fundamental, and cannot be settled by an arbitrary convention which has no roots in an underlying reality. Observers may be trained to agree closely with one another in making very subtle discriminations in the identification of visual patterns (3.2) – akin to the extraordinary skills of the professional wine taster or perfume blender – yet there may be great difficulty in organising such consensual knowledge into a rational map (4.2). Thus, for example, the development of hominid palaeontology is held up for the lack of a well established classification of fossil skulls (6.5) that has more meaning than, say, the geographical location of the finds or the supposed volumes of the brains they once housed.

But here, at least, we may be confident that the discovery of new

---

[1] This difficulty cannot be avoided by setting up a theoretical system in a mathematical formalism satisfying the axioms of 3-valued logic (5.10): all that this could tell us, at the end of a calculation, would be the extent of the uncertainty of the answer due to the initial vagueness of the starting assumptions. This is a salutary exercise for any model-builder in any branch of science, but would always, I suspect, confirm his worst fears, and reinforce the point I am here making.

specimens and the applications of new techniques such as radioactive dating will clarify the situation until the true 'family tree' of man has been disclosed. In other words, *biology*, although seldom considered one of the 'exact' sciences, exhibits a whole hierarchy of very well-defined categorial schemata (3.2, 6.7) ranging from the constituent nucleotides of DNA and proteins, through organelles, cells and organs, to whole organisms, populations and ecological habitats. At each level there is an astonishing degree of identity amongst the elements of a particular class, and quite obvious differences between those of different classes. Anatomically speaking 'the colonel's lady and Judy O'Grady' *are* 'the same beneath the skin', and yet completely distinguishable from horses, jellyfish, or gooseberry bushes. Despite the fact that the instrument of discrimination may be no more 'scientific' than the human eye and brain (3.2), the categories of biology are not merely arbitrary classes with conventional boundaries; they refer to, and arise from, the fantastic natural order that is intrinsic to every aspect of the domain of living things.[2] Extremely powerful, well-validated, logically unambiguous, highly predictive sciences such as molecular and cellular biology, anatomy, physiology, pathology, genetics and evolutionary taxonomy can thus be constructed on firm foundations of sharply defined and rationally ordered categories. It is only within this framework that we can recognize the secondary problems that arise in categorizing such entities as polygenic traits, botanical cultivars, interspecific hybrids, vestigial organs, etc. which seem to cross the boundaries we have learnt to observe and respect.

The social domain does not lack meaningful categories. Anthropology teaches the diversity and ubiquity of institutional and cultural classifications to be found in all human groups (7.9) 'Lord' and 'serf'; 'Brahmin' and 'untouchable'; 'worker', 'peasant' and 'intellectual';

[2] The transcultural consensibility (5.10) of the biosphere is attested by R. Bulmer (1967) *Man*, **2**, 5–25 (reprinted in Douglas 1973) 'Why the cassowary is not a bird'. In New Guinea, the 'Karam show an enormous, detailed and on the whole highly accurate knowledge of natural history, and...though, even with vertebrate animals, their terminal taxa only correspond well in about 60 per cent of the cases with the species recognized by the scientific zoologist, they are nevertheless in general well aware of species differences among larger and more familiar creatures. The general consistency with which, in nature, morphological differences are correlated with differences in habitat, feeding habits, call notes, and other aspects of behaviour is the inevitable starting point for any system of animal classification, at the lowest level.'

It is not inconsistent with our present thesis to note further that 'at the upper level of Karam taxonomy, however...culture takes over and determines the selection of taxonomically significant characters. It is not surprising that the result shows little correspondence either to the taxonomy of the professional zoologist, which reflects the theory of evolution, or, for that matter, to our modern Western folk-taxonomies.'

'millionaire' and 'pauper': these, for example, are categories of the utmost significance which dominate the lives of their respective members. But they are not *sharp* categories. Even in the Hindu caste system, there are innumerable sub-groups and subtle gradations. From the pauper to the millionaire there is the nearly continuous scale of monetary incomes – but even this numerical measure has none of the precision of a length or a mass of a permanent object, and cannot be subdivided indefinitely into distinctly different categories. The same difficulties arise in the classification of individual behaviour; in the study of mental disease, for example, the categorial schemes distinguishing between 'neurosis', 'hysteria', 'psychosis' etc. are as vague and questionable as the specific treatments that are advocated for them.

It is not difficult, of course, to define *formal* categories in the social domain. But these categories are prescribed only by artificial conventions, and need have no significant reality outside the notebooks or computer tapes of the investigator. 'Labour voters', for example, might be defined as 'Persons who voted for official Labour candidates in the 1945 Election'; but did that particular personal decision on that particular day signify a political attitude as invariant as, say, the attitude to child-rearing signalled by the call of the cuckoo. The 'Lords' are, no doubt, 'Hereditary members of the House of Lords'; but are such legally defined persons any more homogeneous as group of human beings than the collection of animals in a farmyard as zoological organisms? The problem of discovering *significant* categories in the social and behavioural sciences is not solved by such methodological practices: the challenge to sociological research is to *demonstrate* that such artifical categories are, indeed, well-defined, stable, consensible and meaningful as elements in a conceptual scheme. The sharpness of the initial classification is soon blurred as this investigation proceeds.

The fact is that we do not have a taxonomic framework for human behaviour, individual or social, in which the categories are both meaningful and distinct. The question for us is not whether such a scheme is in principle unattainable; it is simply whether this essential ground base for a consensual science has actually been found. Without it, we must be wary of any attempt to analyse such topics within the confines of two-valued logic. In every social or behavioural system, there must be elements which cannot be obviously categorized, so that all formal logical implications, truth tables, etc. must be taken with a large pinch of salt for the uncertain, unprovable, 'don't know' classes.

7.2 Categorial imprecision

This is not an absolute bar to consensible discourse, or even some reliable consensual knowledge in this field. But the underlying three-valued logic of argument must not be obscured or ignored. This feature of the social sciences is inherent in the widespread use of *statistical* descriptions and analyses (3.5). Nevertheless, the standards of observational precision, theoretical prediction and validation in these sciences cannot come near to those that are taken for granted in the biological and physical sciences, where sharp, distinct categories of almost identical objects are easily observed.

7.3 The algebra of social experience

Even when quite distinct categories can be defined in the social domain, the mathematical machine (2.4) is not necessarily applicable. The immense power of mathematics in the physical sciences derives from the possibilities of formal manipulations within the domain of theory. Physical quantities or theoretical concepts are represented by symbols, numbers or geometrical figures from which successive algebraic, numerical, geometrical or topological transformations generate quite unsuspected logical relationships. Starting from, say, the hypothetical properties of a model, the applied mathematician arrives at quite specific predictions (2.8) whose precise confirmation validates his whole chain of assumptions and manipulations.

For such intellectual feats to succeed, however, the formal properties of the mathematical symbols must be isomorphous with the empirical relations of the categories they purport to represent in the real world. The language of theory must correctly mirror reality. In physics, of course, this isomorphism is imposed by convention; it is taken for granted that nature is inherently 'mathematical', and the subject matter for research is selected accordingly (2.7). In other branches of science it is necessary to determine the relationships that actually exist between various naturally occurring categories, and to build up an appropriate mathematical theory – if possible.

Consider again the familiar symbolism of chemistry. An 'equation' such as (4.3)

$$Zn + H_2SO_4 \rightleftharpoons ZnSO_4 + H_2$$

could have been written down by an alchemist as a cryptic record of an observation that zinc and sulphuric acid combine to produce hydrogen gas. But the history of chemistry until the early nineteenth century might be described as a search for a formalism from which much more could be 'read' than a mere catalogue of reactions. It was the genius of Dalton to show that symbols representing standard masses of various chemical species had the algebraic virtue of *commutative additivity* – i.e. that these actual weights of zinc etc. could be made to combine in precisely these proportions to give just such and such weights of products and that this result would be the same whether we add the zinc, pellet by pellet, to the acid, or whether we pour the latter drop by drop on to the metal. Of course the symbolism does not tell us *everything* about such reactions, but it can be rearranged to 'predict' that quite a different process, such as

$$ZnSO_4 \rightleftharpoons ZnO + SO_2 + H_2O,$$

if it could happen at all, would give rise to calculable quantities of zinc oxide, sulphur dioxide and water. By virtue of the chemical law of conservation of mass, these symbols obey abstract algebraic axioms of addition and composition that faithfully mirror the relations of the real categories they represent. These axioms, moreover, are not trivial or vacuous, for they permit innumerable theoretical manipulations, transformations, substitutions and recombinations whose outcome can be verified experimentally. Quite apart from their interpretation in terms of physically distinct atoms, the symbols of chemistry constitute a simple *map* (4.2) of much of our knowledge of the subject.

The question is whether any similar abstract formalism has yet been found in the social or behavioural sciences. Do we find categories of psychological or sociological experience whose logic is isomorphous with any manipulable formalism? Even where quite sharply defined categorial schemes have been discovered, they do not seem to conform to any such simple principles as those implicit in elementary algebra or other well-explored and powerful mathematical systems.

The situation is exemplified by the standard convention of representing 'intelligence' by a numerical 'IQ'. Let us concede that the IQ of each individual might be a measurable, invariant quantity, from which one could deduce, for example, the probability of that individual successfully solving a particular type of puzzle. But what meaning could possibly be attached to the arithmetical operation of *adding* IQs? The numerical identity

$$87 + 133 = 110 + 110$$

has no counterpart in the intelligence dimension; it would be nonsense to suppose that a person with IQ = 133 working with a person of IQ = 87 could solve the same problems as two people, each of IQ = 110, working together. The numerical representation of intelligence only satisfies the algebra of *ordinals*, and not of cardinal numbers.

This may seem a trivial point. But whenever the *average* IQ of a group is calculated and quoted as a scientific datum, there is the underlying assumption that the IQs of individuals *can* be aggregated and divided numerically in just this way, just as we might calculate the average energy of the particles of a gas. But there is no 'law of the conservation of intelligence', analogous to the law of conservation of energy, which would legitimate such a procedure. In social psychology, therefore, there is very little hope of deriving interesting, unsuspected, theoretical conclusions by a mathematical analysis of the kind that is so extraordinarily fruitful in the physical science of thermodynamics. Of course there is always the reductionist objective of finding a wonderful new calculus that will break through this barrier (7.5) but this aspiration is not warranted by the knowledge available to us today.

Of all the social sciences, the nearest analogue of physics is presumably *economics*. But the fact that many goods and services have monetary prices and costs that can be aggregated and manipulated arithmetically does not mean that *all* social products and activities can be similarly quantified. The paradoxes of cost-benefit analysis are well known. At the most fundamental level, the exercise of assigning a money value to the good life is fatally frustrated by the simple fact that the value to an individual of his *own* life is essentially infinite – for without it he has nothing anyway – whilst in the long run the worth of any man's life is practically nil – for, as Frederick the Great reminded his soldiers when they hesitated in battle, 'You rascals, do you want to live for ever!'

It is sobering to consider another social and psychological factor of great importance – namely 'information'. Despite all our efforts, this cannot be treated as an economic commodity, for it cannot be assigned any fixed 'value' that does not depend completely on the whole context within which it is being evaluated. Neither the conventions of book-keeping nor any algebraic formalism can represent the realities of a factor that is totally historical and circumstantial in its significance. That is not to say that an industrial entrepreneur or a spymaster does not know the worth of what he is buying; only that this figure is not a quantity that could be assigned invariantly to the message itself thus transferred.

Once more, we must refrain from entering further into a vast subject. The literature on the quantification or other formal representation of social phenomena is beyond summary, and is certainly not oblivious to the above remarks.[3] Nor need we deny that much progress can be made by counting heads, or indoor toilets, or the answers to opinion polls. It is important, nevertheless, to keep in mind that the use of mathematical language is not a goal for its own sake in the natural sciences, but is a highly specialized technique for arriving at a very complete consensus. Any attempt to apply the same intellectual technique to social and behavioural phenomena must have the same purpose, and must therefore maintain comparable criteria of categorial precision and operational fidelity to the realities. By these criteria, unfortunately, most such attempts appear implausible and pretentious and probably induce obfuscation and mental opacity where intended to clarify and persuade.

7.4 *Experimental simplification*

It would be wrong to suppose that these difficulties are peculiar to the social sciences. At first sight, most of the phenomena of nature are as complex and non-logical as those of human behaviour. The problems of discovering consensible categories and consensual patterns in such perverse observables as the stars in the sky, a geographical landscape, a tree, a piece of rock, or a proliferating fungoid growth, have been the perennial problems of science.

These problems have usually been solved by resort to *experiment*. This characteristically scientific 'method' not only produces consensible observational information (3.3); it is also the means by which the material domain is caught in unusually simplified attitudes, where its inner workings are laid bare. Once again, this interference with the natural order is most extreme in the physical sciences, where the most elaborate and extravagant devices are contrived to generate specimens of matter in peculiarly simple, mathematically definable states – perfect crystals, rarefied gases, uniform beams of energetic particles, etc. (2.7). But the same strategy is used throughout biology, where single cells are teased out of the surrounding flesh and studied microscopically, where peculiar regimens of diet are imposed on living organisms, or where vast families of plants or animals are deliberately mated and bred under controlled conditions. The phenomena that are

---

[3] See, for example, J. S. Coleman (1964) *Introduction to Mathematical Sociology* (New York: Free Press of Glencoe) which concedes most of the above points and is very cautious in its claims.

observed in these conditions are, so to speak, quite unnatural, but by eliminating many random influences, and sharpening almost to caricature the observable categories of experience, the experimenter can acquire a logical grasp of the situation, and build and test theoretical models which ought to be applicable under more normal conditions.

This strategy is, of course, widely applied in the social and behavioural sciences. The academic discipline of *experimental psychology* is devoted to the study of the behaviour of individuals in carefully contrived and standardised situations – sitting quietly at a desk, alone in a featureless room, waiting, pen in hand, to make a mark on a piece of paper when a light flashes or a buzzer sounds. Observational consensibility leading to consensual descriptions of typical 'phenomena' is quite capable of being achieved in such experiments. The introduction of social interactions between several 'experimental subjects' in the field of *small-group psychology* (7.7) demands much more stringent precautions to keep the disturbing variables under control, but some progress has been made in discovering more or less reproducible phenomena, such as the tendency of individuals to conform to group opinion.

Nevertheless, one comes away from accounts of such research somewhat sceptical of its ultimate value as a source of deep understanding of human behaviour. The factors that the experimenter endeavours to eliminate by his choice of a homogeneous group of subjects – typically college students in psychology – may be precisely those at issue; for example, is the tendency towards conformity a universal human characteristic, or is it very different amongst the young or the old, very different amongst the young in *relation* to the old, very different amongst Japanese peasants or French intellectuals? These are, of course, all legitimate questions for study by the same methods, but a little knowledge of men and affairs would suggest that the answers would not be simple, uniform, or easy to map and interpret. In other words, human behaviour, whether of individuals or of small groups, is extraordinarily complex, and a human being cannot be divorced from his historical and cultural circumstances as easily as a botanical specimen transplanted into a greenhouse from its natural environment (7.10). This research often contributes to our understanding by falsifying crude or naive generalizations about 'human nature' (7.9). But it has not yet established entirely novel findings that must command unquestionable assent by their reproducibility and consensibility.

What we must also ask is whether the results of such research could *ever* hope to explain real social phenomena. Is the volunteer taking part in a psychological experiment, fully trusting the benevolence and sanity of the research worker, the same person as the bored and fatigued factory worker, the slightly alcoholic salesman, the totally overburdened politician, or the terrified front-line soldier, whose reactions to events he is supposed to be mirroring? Can we honestly believe that the behaviour of very large human societies, involving many thousands of individuals with many institutional loyalties and private interests, must be a larger version of what is observed when a few dozen people are brought together for a few days or a few hours under artificial conditions? It is chastening to note that political systems dominated and terrorized by psychopathic monsters such as Hitler, Stalin, 'Papa Doc', or Idi Amin are not at all uncommon in the real world, and evidently have their own characteristic roots in the human psyche.[4] Such a phenomenon cannot conceivably be 'reduced' to, nor extrapolated from, what could be discovered and reproduced on a small scale by an 'ethical' sociological experiment.

It must be emphasized, once more, that the scientific status of experimental research on human behaviour is not being denied. Many of the problems associated with the self-consciousness of the 'subjects', and the bias of the observer can be largely overcome. But we must ask whether we do, in fact, obtain from this research some truly invaluable insights that we would rather trust than many alternative sources of knowledge (7.10) about the ways of our fellows.

7.5 Hidden variables

The observed behaviour of human beings, whether in the natural life of society or in contrived experimental circumstances, does not seem to conform to simple logical patterns or to mathematical formulae (7.2–7.4). The only remaining hope for a strongly predictive or explanatory theory is to hypothesize hidden mechanisms with relatively simple properties of which the observed behaviour is a complex manifestation. The paradigm of such a theory might be the *atomic hypothesis*, which provided a very complete explanation of the chemical phenomena that were known at the beginning of the nineteenth century, and which was the inspiration of Dalton's symbolic reformulation of this evidence (7.3). This hypothesis had long been one of

[4] See E. Canetti (1973) *Crowds and Power* (London: Penguin).

the familiar *themata* of Western science (6.7), known to all educated persons through the writings of the classical Greek philosophers. The 'atomic revolution' in chemistry was the recognition that from this model could be derived, with complete quantitative precision, the multifarious, infinitely complex facts concerning the combining weights in all chemical reactions. It is noteworthy, moreover, that the *physical* properties of real atoms – their sizes, shapes, masses, charges, mechanical interactions – remained almost unknown for the best part of the nineteenth century, and were not needed by the chemists in their bold exploration of innumerable natural and synthetic compounds. *Genetics* made similar remarkable progress on the basis of Mendel's model of the combination of 'genes', long before these could be shown to consist of strings of 'codons' along the chain-like molecules of DNA. In the natural sciences there is thus a well-established intellectual tradition by which *hidden variables* may be invoked as a putative explanation of complex observed phenomena long before these variables can be given further attributes of 'reality' (6.4) by other investigations.

It is almost impossible for the behavioural scientist to avoid thinking in similar terms. Psychological and sociological theorizing abounds in 'disposition properties'[5] such as 'anomie', 'morale', 'integration', etc. which are not directly observable, not defined operationally, and seldom deducible by strong logical inference from the actual evidence. It is very difficult to cast a virgin eye upon the phenomena of social life, uninfluenced by past interpretations or models. Indeed, human behaviour would appear entirely meaningless or chaotic to an observer unfamiliar with its complexities, and not already provided with some 'picture' (4.3) of the social world drawn from his own experience or from the reports of others (7.9). This is the fundamental problem of the anthropologist.

In fact, it is quite easy to conceive quite interesting 'maps' of human behaviour and to convince oneself that they must be true (4.2). But having by some means or another hypothesized an underlying model mechanism to explain what is going on, how can we persuade *others* that this must, indeed, be the correct explanation.

Unfortunately, human behaviour is always so complex and varied that we can seldom make a sharply confirmable (or disconfirmable) prediction from the model. At best, the chain of inference can only be tested *statistically* (3.5); the hidden machinery of the model produces

[5] Coleman, *Introduction to Mathematical Sociology*, p. 75.

no more than 'tendencies' in particular directions, with few strictly deterministic outcomes. The logical path from observable data with much uncontrolled variance back to the supposedly simpler and more mechanically related 'dispositions' or 'forces' is gravely uncertain.

A quite modest and unassuming example of this difficulty is the use of *factor analysis* to explain the results of aptitude tests, intelligence tests, etc. The underlying hypothesis – that intellectual and sensori-motor performance in a variety of circumstances is governed by a small number of 'factors' acting almost independently and combining almost linearly – is not fundamentally implausible. The objection to this procedure is simply the impossibility of demonstrating from the data that this sort of interpretation is as 'necessary' (i.e. as thoroughly convincing) as, say, the atomic model of chemistry, or the gene model in genetics. It must never be forgotten that statistical argument never actually *confirms* a hypothesis; it can only tell us, more or less loosely, whether the data are *consistent* with our theoretical assumptions.

The analogy with physical phenomena is particularly misleading in this case. It is perfectly true that important physical discoveries have been made by inspired scrutiny of apparently featureless data. The first evidence for the existence of *pulsars*, for example, came from the observation of regular pulses amongst the otherwise, random radiation received by a radio telescope from outer space (3.6). The claim of Joseph Weber to have observed gravitational waves (3.5) is of the same form of a supposed *signal* amidst a great deal of *noise*. In both these cases, however, the received data can be processed by exact techniques, many of the sources of noise can themselves be calibrated, and a diversity of alternative instruments can be constructed to detect the 'signal' by other means. These conditions are seldom to be found in cases where factor analysis is the only means of access to the model variables. Reliable scientific knowledge must pass the most rigorous tests of scepticism and attempted falsification. The consensus principle demands that the evidences and arguments for a proposed theoretical interpretation must eventually be overwhelming. Statistical inferences from data with large and uncontrollable variances are often 'interesting' but seldom convincing. Thus, for example, the Sternglass data (3.5) could not be altogether disregarded; but they are quite inadequate to *prove* his hypothesis of a direct causal connection between nuclear explosions and infant mortality. 'Hidden variable' models of social behaviour, though often immensely valuable as a stimulus to further thought and further research, can seldom transcend the three-valued logic of the observational categories (7.2) they are supposed to explain.

More harshly, it might be said that the behavioural sciences are cluttered with innumerable half-articulated speculative models of this kind that have never been subjected to critical validation. Standards of theory construction and confirmation have seldom been sufficiently high to distinguish clearly between what is well established, what is essentially conjectural, and what has been thoroughly disconfirmed (6.5). Theoretical 'sketches' abound, but there are very few reliable 'maps'. Many of the 'pictures' in the minds of research workers and practitioners are sheer fantasy, contradictory in themselves and having no basis in reality. Such a situation is, of course, deplorable; but it reflects the enormous difficulties of defining consensible observations and discovering consensual theories to explain them.

7.6   *Models, toys and games*

Yet the dream of making a working model of social behaviour continues to grip the scientific imagination. It is conceded that such a model cannot be so simple as those conceived by past social theorists, but should nevertheless (so we are told) yield to the number-crunching power of a high-speed electronic computer (6.7). The aspiration of *general systems analysis* is to provide an instrument by which economists, businessmen, administrators and politicians can make sound practical predictions. It has shown its value in the efficient management of material objects, such as the supply of components for the manufacture of motor cars, or the targetting of nuclear weapons on hostile cities; eventually it might help to reach less tangible goals such as national social development or personal health.

We must regard these latter claims with considerable scepticism. There is, as yet, no evidence that any progress at all has been made by such means. Even the most elaborate economic models have failed to solve the problems of inflation, unemployment, exchange rates, etc. However detailed the mathematical computation, however many factors are supposedly taken into account, such an analysis is completely at the mercy of the assumptions that are made in setting it up. These assumptions, both as to the boundary conditions and constitutive equations, are subject to all the doubts and uncertainties discussed above (7.2–7.5). In many cases, the crudity and unreality of these assumptions are hidden under a mass of mathematical formulae, thus mercifully rendering a dubious argument completely opaque. Even when there is no claim to quantitative verisimillitude, the model usually contains doubtful qualitative relations between vague categories (7.2) or hypothesized hidden variables (7.5) which have never been shown to have any operational invariance. Such work must therefore be approached with a very cool eye. There is no reason why a piece of formal reasoning should be regarded as persuasive just because it is impenetrably complicated, when it does not satisfy in detail the elementary canons of scientific credibility.

Indeed, what is often so puzzling about this sort of work is the status that is being claimed for the outcome of the calculations. The model can never be said to be so well founded that its predictions could be taken as seriously as those, say, of the performance of a newly designed aircraft. Nor are arrangements made to collect sufficient material evidence to confirm the predictions in detail, hence validating the assumptions of the model. On the other hand, the model itself is usually much too complicated to exemplify a general principle or to demonstrate a hypothetical phenomenon. It is thus difficult to decide what has been added to the archive of scientific knowledge by such investigations.

In the physical sciences, theoretical progress often follows on the invention of 'toys'[6] – models with well-defined properties not precisely representative of any real system, but simple enough to be studied mathematically in depth and detail. Typical examples are the familiar model of a 'perfect gas' as an assembly of massive point particles or Bohr's 'liquid drop' model of the atomic nucleus (2.5, 4.2). Since these models are conceived hypothetically, and need have their being only in the noetic domain (5.5), they may be assigned any logical properties that are convenient. For mathematical analysis, sharp two-valued logic (2.6) is obviously preferable.

As he 'toys' with such a model, the theoretical physicist deduces a number of qualitative or semi-quantitative features or behavioural properties that may perhaps appear similar to the observed properties of real systems. Thus, the 'perfect gas' toy demonstrates many of the phenomena associated with real gases, such as Boyle's law, and is thus a conceptual basis for a more realistic kinetic theory of gases, leading on to the much more sophisticated mathematical theory of statistical mechanics. From the 'liquid drop' model, Bohr was able to interpret the mysterious phenomenon of nuclear fission without the necessity

[6] J. M. Ziman, 'Mathematical models and physical toys' (1965) *Nature*, **206**, 1187–92.

of accounting in detail for the many-body forces involving the several hundred nucleons in an actual nucleus.

The essential point of such research is that the deduction of these properties of the 'toy' can be made a rigorous mathematical problem, whose methods are fully consensible and whose final results are logically consensual. They may not eventually turn out to be relevant to observation or experience, but at least some well-defined, well-established propositions can be added to the scientific archive. For this reason, much of the work of the professional theoretical physicist is devoted to the elucidation of the subtle mathematical properties of a range of currently fashionable 'toy' theories, which are not actually modelled closely on any particular real physical system.

The careful study of such *synthetic theories*[7] also has its value in the social sciences. This, for example, is the epistemological strategy of the *Theory of Games*, introduced by von Neuman and Morgenstern as a contribution to the theory of economic behaviour. In the first instance, it is not to be questioned whether the axioms and definitions of 'players', 'game', 'winnings', etc. correspond accurately to any economic reality.[8] The purpose of the investigation is simply to discover the inner structure and latent properties of the toy as axiomatically defined. Some surprising theorems can then be proved – for example that 'bluffing' at a game like poker is not a form of psychological trickery or dishonest play but a rational strategy that maximizes the potential winnings of the player. In other words, a phenomenon of uncertain significance in the real world derives directly from certain simple structural features of the system.

The danger is that enormous efforts can be made towards the displaced goal of solving more and more elaborate mathematical puzzles that have less and less relevance to any conceivable reality. The art of mathematical physics (and, presumably, of mathematical economics) is to invent 'toys' that have some essential resemblance to what may be imagined in the material domain and that are also mathematically tractable. This is a very difficult thing to do: but it is inherently more creative than the postulation of theoretical systems that are too vague, too loosely articulated, or too ponderously conforming with known detail, to be capable of generating qualitative novelty or falsifiable predictions. Scientific method, in the search for consensus, must provide critical issues on which theories may stand

[7] Coleman, *Introduction to Mathematical Sociology*, p. 41.
[8] See e.g. R. Handy (1964) *Methodology of the Behavioural Sciences* (Springfield, Illinois: Thomas).

or fall; these often arise from concentrated research on well-posed synthetic theories, with two-valued logic, in the noetic 'Third World'.

7.7 Simulations

Economic games acquire a semblance of realism through the appearance of a quantitative variable, 'money', whose formal properties can be assumed to be those of elementary arithmetic. It is much more difficult to construct analytical systems involving psychological or social factors such as 'prestige' or 'political bias' or 'anomie'. The logical properties that may be assigned to such 'variables' are so vague, or so implausible (7.3), that a formal symbolic analysis tells us little more than would appear obvious to a thoughtful participant or observer. To point out, for example, that a conflict of human wills may be mathematically analogous to a mechanical device with two extreme states of stable equilibrium opens the mind to new ways of picturing the situation, but does not really uncover a means of resolving such conflicts. It is seldom profitable to turn the mathematical machine (2.4) on to such a system in the hope of gaining deeper insight.

The really interesting characteristics of human behaviour are much more subtle and convoluted (7.4) than those that can be modelled by a quasi-linear, quasi-deterministic mathematical representation. To demonstrate these characteristics under controlled, reproducible, consensible conditions, we may need the aid of a different type of 'toy' – the playing out of a *social simulation*.

Such toys, indeed, are as old as the hills. In ancient *board games*, such as *Chess* and *Go*, the players take on the social roles of opposing generals. Despite the simplicity of the rules, the play itself acquires immense intellectual depth, and evinces behaviour that has a striking resemblance to the behaviour of real men in war-like conflict. Strategic principles that might have come from von Clausewitz[9] – 'surprise', 'concentration of forces', 'deployment of reserves', 'exploitation of opportunities' – are highly significant in successful play. In other words, characteristic forms of 'social' or 'psychological' behaviour are generated within the sharp categories and rigid rules of the game, and can thus be seen to be necessitated by those rules and not by, say, the more complex and confused circumstances of actual war.

This point is emphasized by recent achievements in programming a computer to play chess to a relatively high standard. This does not

[9] *On War* (1832; English translation 1908: republished 1968 by Penguin Books, London).

mean that the infinite complexities of a real military decision could be rationalized and confided to a quasi-infinite computer. But it does show that the intuitive heuristic principles of strategy under precisely defined and consensible conditions can arise naturally from the 'logic of the situation' and do not necessarily contain 'psychological' elements. Yet a board game is very far from a simple system. The topological conditions of adjacency, occupancy, territoriality, etc. that are implicit in a game like *Go* are easily grasped visually (3.2, 5.4) but are quite complicated to programme into a computer. Considered as a formal code, these conditions would look quite arbitrary and meaningless without reference to the map relations that underlie them. The game is 'realizable' as a consistent model system (2.5) and as a social simulation because it is dominated by the same geographical logic that has traditionally dominated the art of war itself.

In recent years, such games have been enormously elaborated, quite beyond the reach of robotic participation.[10] In *Inter-Nation Simulation* for example, various players pretend to be the political leaders of great nations, sending diplomatic communications to one another, making 'economic', 'political' and 'military' decisions, combining as allies, etc. Getting away from geographical relations altogether, there are games such as *Simsoc* which are designed to simulate the social relations and interactions within a schematized and simplified social system.

Simulations where the action at each node of the network is not decided by a mathematical formula but by a person temporarily 'playing' a social role, are not of course deterministic, but are widely employed for training social workers, staff officers, diplomats, business executives and others. They are also valued as sources of insight into characteristic forms of human behaviour under more or less realistic conditions. In this intellectual domain, they may be regarded as equivalent to the 'toys' and 'games' of the theoretical physicist or economist; from them it is hoped that the mechanisms and conditions for various social phenomena may be elucidated.

Strictly speaking, however, a social simulation under controlled conditions is an open-ended small-group experiment (7.4). The underlying assumption must be that 'one man is as good as another' as a participant in the drama. If the game is to have some semblance of realism, then Bill Bloggs, who has drawn the role of 'leader of Redland' must be assumed to be in some senses equivalent to a Nikita Krushchev or a Leonid Brezhnev as head of the Soviet Union in a

[10] See e.g. M. Inbar and C. S. Stoll (1962; eds) *Simulation and Gaming in Social Science* (New York: The Free Press).

diplomatic crisis. Various chance factors will, of course, influence the actual outcome of each round of play, but if the protocol has been well designed there should be a recognizable uniformity in the record of many rounds of play.

Such an assumption is not entirely without substance. The crude equivalence of human beings in their powers of thought and feeling (7.8) must be the foundation of any generalizations concerning social behaviour. But the objections to simplified experimentation as a source of reliable knowledge in the social domain (7.4) apply with much the same force to the results obtained by social simulations even though these cover a wider range of more complex psychological and social phenomena.

7.8 Humanistic intersubjectivity

Up to this point, we have taken no account of the part played by the scientific observer in the study of behavioural phenomena. He is regarded as a camera, or a robotic note-taker, looking on and recording what he sees with the same objectivity as if he were in front of a bank of instrument dials or looking down a microscope at a colony of amoebae.

In practice, however, the psychologist, sociologist or anthropologist avails himself of the opportunity of communicating with the subjects of his study. He not only observes *what* they do, but often asks them *why* they behave thus. In a simulation game, for example, he may be told 'it was *blackmail*', or 'I had to show that we couldn't be *bullied*', or 'I felt terrible – letting my *allies* down so badly'. Now of course the observer must be very careful not to influence events by his own comments, or become too emotionally involved with the subjects of his experiment. But this does not mean that such communications are meaningless or irrelevant; on the contrary, they may provide the most valuable clues to the interpretation of what is observed. It is precisely this access to the mental domains of the human actors that allows us an understanding of quite complex human social behaviour which would be almost unintelligible if observed amongst chimpanzees or dolphins.

What is more – and here we approach the central theme of this chapter – *such communications are similarly intelligible, and would carry much the same message, to any other scientific observer under similar conditions*. As a body of public knowledge, science depends on the consensibility of the observations made by scientists, and the eventual consensus achieved by the exchange of messages describing these observations (1.4). Valid scientific observations include not only quantitative instrumental measurements, but also records of visually recognized 'patterns' (3.2) which are consensual for most normal human beings (5.9). An essential foundation for reliable *social* knowledge is a further degree of consensibility concerning the *personal* meaning of social actions, as expressed by the actor. The counterpart of the intersubjectivity of visual pattern recognition for the material domain is the *humanistic intersubjectivity* of motive, emotion, reflection, intention, etc. for the social or psychological domain.[11]

Thus, for example, the meaning of *blackmail* or of *being bullied* is well understood by almost any adult person from his own experience or from drama or literature. Recognition and discrimination of psychological entities of this kind is as widely shared, and as necessary for the construction of a science of behaviour, as the visual recognition of cats, dogs and kangaroos for the construction of a science of zoology. If such a science is to get under way at all, it must begin with observational reports firmly based on this commonplace consensibility.

This is not to deny the value of rational analysis of psychological situations, reducing complex entities as far as possible to simpler elements: in just the same way, the natural scientist attempts to simplify and analyse his visual observations to improve the consensuality of their interpretation. Nor need we rule out the possibility, in some dim and distant future, of constructing a Turing robot programmed algorithmically to behave 'just like a man' in social interaction – just as we may aspire to construct a universal pattern-recognition machine that matches human capacities of perception (5.3). It is sufficient to remark that these idealizations have not yet been realized, that our human powers both of psychological understanding and of visual perception are still 'extra-logical' – and that we cannot delay the scientific exploration of the natural and social worlds until we think we have solved this problem in practice as well as in principle.

The fundamental importance of *empathy* in the social sciences has often been emphasized. The names of Dilthey and Collingwood are

---

[11] As remarked, for example, by J. Heritage (1974) in *Reconstructing Social Psychology* edited by N. Armistead (London: Penguin) 'The accountability of actions in turn depends on *shared* access to common rationales and procedures through which the 'sense' and 'order' of events can be made intersubjectively available'. Margaret Mead (1976, *Science*, **191**, 903) in her address as President of the American Association for the Advancement of Science, states this principle perfectly 'our knowledge of ourselves and of the universe within which we live comes not from a single source but, instead, from two sources – from our capacity to explore human responses to events in which we and others participate through intro-spection and empathy, as well as from our capacity to make objective observations on physical and animate nature.'

associated with this view as a basis for a philosophy of history – but even the great von Ranke, who strove so hard to eliminate the 'personal' from his writing of history, derived much of his intellectual authority from his sympathetic understanding.[12] Whether or not so humanistic a faculty could or should form the corner-stone of this and other branches of the study of human behaviour, there is general agreement that it is an indispensable source of knowledge.[13] In every historical, sociological, anthropological, or psychological situation there are, of course, many elements and factors capable of rigorous logical analysis according to rational principles,[14] but factors of emotion and human value cannot be brought into public discussion without an appeal to the empathic authority of the common humanity of actors and observers.[15]

Of course if one tries to take this argument too far, to insist that observer and actor must share precisely the same inner world of thoughts and emotions,[16] then one runs into a contradiction. But such total psychic identification is not called for in our model of science. The empathy of the observer with the actor need be no greater than is consensible in the community of scientific observers, who must convey to one another their observations and theories. What they share is no more than a variety of analogous personal experiences and a language in which something can be said about inner states of consciousness.[17] The political mystery, the psychological incompre-

---

[12] This point is made by E. Cassirer (1956) in *The Problem of Knowledge* (New Haven: Yale University Press) p. 237.

[13] Thus, E. Nagel (1961) in *The Structure of Science: Problems in the Logic of Scientific Explanation* (London: Routledge & Kegan Paul) p. 483 asks whether the social scientist can 'account for men's actions unless he has experienced in his own person the psychic states he imputes to them, or unless he can successfully recreate such states in imagination.'

[14] Popper, *Objective Knowledge*, p. 187.

[15] J. Watkins (1970) in *Explanation in the Behavioural Sciences* edited by R. Borger and E. Cioffi (Cambridge University Press) p. 147, sums up thus: 'All this adds up to a thorough critique of any attempt to put actual decision making on a formal analytical basis. But behind it lies the unspoken assumption that rationality is (a) conscious in the agent and (b) sympathetically or intersubjectively 'available' to the person who attempts to repeat the inner monologue implied by evidence about the action'.

[16] Cf. E. Nagel, *The Structure of Science*, p. 483, 'Must a psychiatrist be at least partly demented if he is to be competent for studying the mentally ill. Is a historian incapable of explaining the careers and social changes effected by men like Hitler unless he can recapture in imagination the frenzied hatreds that may have animated such an individual?'

[17] A. Schutz (1967) in *Collected papers: I the Problem of Social Reality* (The Hague: Martinus Nijhoff) makes the essential point 'the common-sense knowledge of everyday life is from the outset...structurally socialized;...it is based on the fundamental idealization that if I were to change places with my fellow man I would experience the same sector of the world in substantially the same perspectives as he does, our particular biographical circumstances becoming for all practical purposes at hand irrelevant.'

hensibility, the historical monstrosity of a Hitler or a Stalin lies in his exclusion from the empathic consensus: we cannot hope to understand a creature whose inner life is quite so alien to almost all of us.

7.9 *Origins of empathy*

What are the sources and the limitations of intersubjective consensibility in the psychic domain? Such an investigation closely parallels our earlier study (5.9) of consensibility of verbal description and visual perception. But these new questions are much more complex and ill-defined than those concerned with physical cognition and can only be answered very tentatively and schematically.

Empathy must surely derive from both the inborn traits of the individual and the social environment in which these traits are developed and expressed. In the human species, there is relative uniformity of physiological structures and mechanisms associated with both thought and emotion. Just as we all have eyes, and brains that can see through them, so we seem to have characteristic patterns of feeling that are governed by specific neural loci or activities. It would be impossible to understand the affective consequences of certain operations on the brain, or of various types of drug (such as alcohol), if this were not so. A general 'emotional predisposition' that reacts to the experiences of life in a characteristic way seems as much part of the human birthright as the 'perceptual predisposition' that is similarly developed into the rational faculty of every normal adult.

At the same time, each child is born and grows up in a social world whose patterns and situations are not so grossly different the world over. 'Hatch, match and dispatch'; child and parent; man and woman; leader and follower – these are the invariable circumstances of living, from Newfoundland to New Guinea. It is important to recognize the underlying uniformity of the *consensus gentium* before we are lost in the diversity of its manifestations.[18] To our subtly discriminating eyes and ears there is a marvellous variety in the forms of social life, but we have only to turn to science fiction[19] to realize the very narrow range

[18] Even though C. Geertz (1973) in *The Interpretation of Cultures* (p. 43) remarks: 'the notion that the essence of what it means to be human is most clearly revealed in those features of human culture that are universal rather than those that are distinctive of this people or that is a prejudice we are not necessarily obliged to share'. As an anthropologist he is bound to be conscious of the fact that 'what men are, above all things, is various'.

[19] For example, Ursula Le Guin (1969) in the *Left Hand of Darkness* (London: MacDonald) who imagines in vivid biological, psychological and sociological detail a race of 'human' hermaphrodites.

Social knowledge*

within which this apparent diversity proliferates. From the very depths
of the past, there has been an interactive evolution of the biological,
psychological, social and cultural characteristics of mankind. These
characteristics are adaptively fused into modes of action that we refer
to, in general as human nature. This, we know is not 'the same the
world over' (7.4) but it is surprisingly uniform in all sorts of quite
reliable ways.

As the child develops physically and mentally it thus acquires a sense
of *social* reality comparable with the reality of the material world (5.8).
Both through individual activity and by verbal interaction with others,
it comes to emotional and intellectual consciousness.[20] An important
part in this process is the introjection of the view of the 'other' into
oneself; but the details of this development are not known. All that
can be said with confidence is that the mental domain of the individual
is not after all a purely private and personal region; it is largely
furnished with what comes to it from the culture in which that person
is brought up.[21]

In this process the categorizing and conceptualizing powers of the
human mind (5.4) play their part. The relationships between people,
by kinship or by social function, are given order and meaning. Social
action, like sensorimotor action, has its own inherent logic, which
moulds the language in which it is described, and the categories into
which it is divided. Social knowledge, like our knowledge of the
material world, is automatically and inevitably abstracted into a 'map',
derived from the noetic domain but internalized within the mental
domain of each individual.[22]

It would be wrong to suppose that these internalized 'pictures' or
'maps' (4.3) of the social world are somehow less 'real' than those of
the material domains surveyed by the natural sciences. Feelings, of love
and hate; motives, of pride or envy; relationships, of friendship or

[20] Luria (*The Working Brain*, p. 246) makes this quite general 'Vygotsky introduced into
psychology the concept that the sources of voluntary movement and action lie not
within the organism, not in the direct influence of past experience, but in *man's social
history*; in that work activity in society which marks the origin of human history, and
in that communication between child and adult which was the basis of voluntary
movement and purposive action in ontogeny.'
[21] Geertz (*The Interpretation of Cultures*, pp. 81–8) puts this very clearly: 'Not only ideas
but emotions are cultural artefacts in men...human thinking is primarily an overt
act conducted in terms of the objective materials of the common culture, and only
secondarily a private matter'.
[22] I. C. Jarvie (*Concepts and Society*, pp. 161–72) discusses this map metaphor at length.
'People living in society have to find their way around in it...they construct in their
minds a conceptual map of the society and its features, of their own location among
them, of the possible paths which will lead to their goals and the hazards along each
path...[The] individual's maps are to some extent – but not fully – coordinated with
each other...[and] on occasion – even if not always – individuals can test for the truth
or falsity of their maps'.

competition; institutional affiliations; historical traditions; obligations; duties; social roles; and occupational categories; these determine our every action, and are built firmly into our conscious thoughts and unconscious mind. So strong, indeed, is this sense of social reality that it may be the original source of many of the themata (6.7) that are applied in the natural sciences to physical and biological phenomena.[23] It may be, indeed, that logic itself, the archetype of extra-human categorial necessity (2.3), does not derive so much from sensorimotor experience (5.8) as from the verbalization of family and clan relationships introjected in infancy.[24] The thesis that all intersubjective communication must 'in the final analysis' be anthropomorphic is at least plausible, if too general to be proved or falsified. It is easy to be over-impressed by the superior sharpness of the primary categories of the material domain as in physics (2.7), and to underestimate the consensibility of the much vaguer categories of social and psychological experience.[25]

[23] As suggested in a famous passage by E. Durkheim: 'The contents [of the categories of space, time, totality, causality] are the different aspects of the social being: the category of class was at first indistinct from the concept of the human group; it is the rhythm of social life which is at the basis of the category of time; the territory occupied by the society furnished the material for the category of space; it is the collective force which was the prototype of the concept of efficient force, an essential element in the category of causality'. *Les Formes Elementaires de la Vie Réligieuse* (1968: Paris: Presses Universitaires de France) p. 628.

[24] See e.g. E. Durkheim and M. Mauss (1903) *Primitive Classification* (1963: London: Routledge & Kegan Paul).

[25] This point is made very clearly by J. C. Eccles (1964) in a paper on 'the neuro-physiological basis of experience' published in *The Critical Approach to Science and Philosophy*, edited by M. Bunge (New York: Free Press of Glencoe): 'The illusory nature of the objective-subjective dichotomy of experience is further illustrated by what might be called a spectrum of perceptual experience.

1. The appearance of an object can be confirmed by touching it, and in the same manner can be sensed by other observers, the perception of the object thus achieving public status.

2. The pinprick of a finger can be witnessed by an observer as well as by the subject, but the pain is private to the subject. However, each observer can perform a similar experiment on himself and report his observation of pain, which in this way is shared and so achieves a public status.

3. The dull pain or ache of visceral origin cannot be readily duplicated in another observer, yet clinical investigations have provided a wealth of evidence on the pains characteristic of the visceral diseases, and even on referred pains, so that reports of visceral pain achieve indirectly a kind of public status. Similar considerations apply to such sensations as thirst or hunger.

4. Unlike the preceding three examples, mental pain or anguish is not a consequence of stimulation of receptor organs; yet again a kind of public status can be given to such purely private experiences for there is a measure of agreement in the reports of subjects so afflicted. Similar considerations apply to other emotional experiences such as anger, joy, delight in beauty, awe, and fear.

5. The experiences of dreams and memories are even more uniquely private, belonging as they do even more exclusively to the realm of inner experience; yet again a kind of public status is established by the wealth of communication that there is between observers.'

7.10   The limitations of a science of society

Without empathy, there could be no coherent social order, no communication between men. But behavioural consensibility is incomplete and defective by comparison with the consensibilities of logical thought and perception. We are all deeply immersed in *particular* cultures, whose conceptions of social reality – even of social 'commonsense' – are diverse and eccentric. Every coherent culture maps itself theoretically (4.3) in its noetic domain, but the concordance of such maps is very imperfect. Despite the endeavours of the generalising sociologist, contradictions of world views and ideologies cannot be overcome in the minds and hearts of their respective adherents. Social behaviour is governed, in the large, by introjected individual and corporate *values* whose priorities cannot be decided by overriding criteria acceptable to all men. Such values as those enshrined in the great religions and folk traditions of mankind are not beyond discussion in a consensible language derived from more immediate and mundane experience: the terrible story of Oedipus and the sympathetic Parable of the Good Samaritan must have their counterparts in many other systems of moral tale and myth. But they cannot be reduced to a consensual scheme of general laws comparable, say, with Newtonian mechanics or Mendelian genetics.

This is the domain where the sociologists of knowledge (5.10) correctly emphasize the inevitability of *cultural relativism* (6.4). Argument and counter-argument as to the role of material interest, of class and racial bias, of prejudice and *ideology*, merely emphasize this diversity. In a world that encompasses such contrasts of 'social reality' as the lives of a Texan millionaire, of a Japanese factory worker, of a Nigerian soldier and of a Chinese farmer, the search for a consensual basis for a science of politics or economics appears fruitless.[26] Claims are made for such 'sciences' by their partisans, but they have not proved themselves by reference to the evidence or in application.

Indeed, it would be wrong to suppose that every member of a given culture pictures his social world according to the same map (4.3). The experiences by which each individual learns his way in life are infinitely varied, so that there is great variance in the supposed consensus scheme.[27] Men and women are not absolutely identical and indistin-

[26] John Kennedy's famous assertion 'I too am a Berliner' was good political rhetoric, but sociologically untenable!

[27] W. Goodenough (1974) in *Science*, **186**, 435, remarks, in his review of Geertz (*The Interpretation of Cultures*): 'The narrower the range of variance in the individual conceptions adults have of their group's culture, the easier it is for newcomers to the group, be they children or visiting anthropologists, to create workable conceptions of their own of the group's culture'.

guishable 'particles' like electrons, nor can they be treated as if they were 'as like as peas in a pod'. The more we study people in the supposed uniformity of a mass society, the less plausibly can we assign them to the stereotyped categories of value judgement or interactive behaviour supposedly characteristic of their cultural milieu. Such stereotypes, the highest common factor of immensely diversified sets, are practically useless as guides to action.[28]

It may be, indeed, that science itself, founded deliberately on the consensus principle, has become the central, minimum variance culture of modern society (5.9). As traditional ethical and behavioural systems disintegrate, leaving fragments that diversify and mutate endlessly, only the principles of material rationality and technical competence remain and cohere. The phenomena of *scientism* (2.7) and *parascientism* (6.6) arise as people attribute powers of social, moral, behavioural consensibility to a body of knowledge whose domain of validity is really limited to a much narrower range. In other words, the 'map' of natural science becomes itself a paradigm (4.4) apparently covering a much broader domain of reality than has been strictly surveyed. This paradigm, introjected by formal education into the minds of us all (5.9), has transformed our individual pictures of social reality, giving them spurious coherence and simplicity.[29] It may require considerable personal experience of life, and a sceptical attitude towards many of our own presuppositions, before we can see our own psychological and social world as a 'many-splendoured thing' that cannot be so easily reduced to mechanical or biological themata.

In this search for 'reliable knowledge' concerning human behaviour we thus arrive at a rather disappointing conclusion. There are grave obstacles of categorial vagueness (7.2), operational variation (7.3), experimental irrelevance (7.4), theoretical unprovability (7.5) and cultural relativism in the way of establishing a general 'science of society'.

[28] I. C. Jarvie (*Concepts and Society*) makes this clear: 'The social sciences would appear to be in the awkward position of trying to stand outside the point of view of any one individual and drawing maps of society which only show features on which there is some general agreement, where people's maps overlap'. In his view, the problem is incorrigible: 'But just as no map is useful for all purposes, no social map can correctly represent the whole truth about the social world. The set of all possible useful maps of either the social or the physical world is infinite. And so the search for the perfect map is a dream, an essentialist wild goose chase.'

[29] In the extreme, society itself has to be transformed to realize the paradigm in which it is perceived. H. R. Harré in *Determinants and Controls of Scientific Development* edited by K. D. Knorr, H. Strasser and H. G. Zilian (1975: Dordrecht: Reidel) p. 272, argues that 'society and the institutions within a society are not to be conceived as independent existents of which we conceive icons. Rather they *are* icons which are described in explanations of certain problematic situations.' By what criteria are we to choose between the view that (p. 278) 'societies are conceived according to the same set of options that are present in natural science in terms of which icons of physical reality are conceived', and the Durheimian view (7.9) of which it is a complete inversion?

The conjectural models, the conceptual schemes, the hidden variables, that are often postulated, in perfect good faith, by social scientists have not received consensual status, and do not constitute a working theoretical basis for social action. This is not to deny the value of sociological and anthropological research, which uncovers such richness and variety, such poverty and uniformity, in the lives of all human beings. In default of such research, we should live entirely at the mercy of our own follies, misconceptions and deceptions as to the merest facts of the matter. But interpretative schemes that purport to go deep beneath the surface of social life, to uncover quite unsuspected forces, or to assert necessities that are by no means evident, must be treated with extreme scepticism. The scientific expertise of the aircraft designer or the heart surgeon is to be respected; claims to be able to 'engineer' a social system or to 'doctor' a faltering economy have no such justification in practice or in principle.

And yet it is our familiar experience that psychological and social phenomena are by no means chaotic. In our daily lives we rely heavily upon the predictable behaviour of other people: when trains do not run to time, a mechanical failure is a more common cause than a drunken driver or an erring signalman. Great political, industrial and military organisations can be created and directed by perceptive and shrewd operators. To deny any element of successful calculation in such operations, to assert the complete unreality of the conceptual schemes on which they are founded, is to repudiate the rhetorical force of prediction on which science itself is based (2.8). We must not forget that 'men have acquired through trial and error many unformulated habits of thought which embody sound principles of non-demonstrable reasoning'.[30] Social commonsense[31] is a very satisfactory guide to social action: 'Everyone is, in a certain sense, a fairly competent social scientist and *we must not treat his (or her) theory about the social world and his place in it with contempt.*[32]

Perhaps what we should be asking is whether even this sort of social knowledge can be collected according to agreed principles, communicated consensibly within a scientific community dedicated to public standards of rational proof (6.4), and mapped consensually for the academic archive.[33] The elements of rationality that we intuitively

[30] E. Nagel, *The Structure of Science*, p. 592.
[31] Which, as J. Heritage in *Reconstructing Social Psychology*, edited by Armistead, p. 278, remarks, 'is the first thing we know'.
[32] R. Harré in *Reconstructing Social Psychology*, edited by Armistead, p. 244.
[33] In the spirit suggested by Jarvie (*Concepts and Society*) 'the difference in depth between the maps of ordinary chaps and the maps of sociological chaps is not yet very great. Social scientists...include maps of other chaps'.

recognize in social commonsense – whether in our own familiar culture or in a more alien context – are so permeated with the inexpressible consensibilities of emotional empathy that they seem to evaporate as we attempt to capture them in permanent form. The principles by which effective social action is justified in practice sound too much like 'rationalizations' whose virtue is plausibility rather than logical necessity. What we live by in social life is not a series of scientific laws, but *maxims*,[34] whose lack of consistency and consensuality is irrelevant to their practical value.

The challenge to the behavioural sciences does not come from physics (7.2 'Take us to your Newton') but from the *humanities*. In the search for reliable knowledge about, say, the psychology of sexual relations, do we turn to a book expounding the evidence for 'exchange', 'reward' and 'balance' models of love?[35] Or do we read again our *Anna Karenina*, or *Madame Bovary*, or *Pride and Prejudice*, or Proust, or Saul Bellow, or Patrick White? The novelist, with his sensible ear and discriminating eye, articulates the universal elements in our emotional lives, and teaches us more about mankind than any formal theory. Does the intellectually inclined politician immerse himself in the abstractions of Karl Mannheim, Talcott Parsons – or even Karl Marx? He enlarges his political experience vicariously, he rehearses in anticipation the dramas of his own future, by studying history, by reading and writing biographies, both for pleasure and for profit.

And of all the social and psychological activities of man, few are so subtle, so complex, so demanding of critical judgement, imagination, courage and intuition than the pursuit of knowledge itself. That is why we cannot learn the art of research by reference to the formal philosophies, sociologies and psychologies of science. Epistemology – the assessment of organised knowledge – is a skill acquired by experience, both of particular disciplines and of life itself. We read our Popper and our Kuhn, our Merton and our Polanyi, not for rules and laws and formulae and proofs, but for maxims, and insights, and

---

[34] M. Polanyi, *Personal Knowledge*, p. 31. 'Maxims are rules, the correct application of which is part of the art which they govern. The true maxims of golfing and poetry increase our insight into golfing or poetry, and may even give valuable guidance to golfers and poets; but these maxims would instantly condemn themselves to absurdity if they tried to replace the golfer's skill or the poet's art. Maxims cannot be understood, still less applied by anyone not already possessing a good practical knowledge of the art. They derive their interest from our appreciation of the art, and cannot themselves either replace or establish that appreciation.'

[35] E.g. *Theories of Attraction and Love*, edited by B. I. Murstein (1971: New York: Springer).

understanding. We also turn to the makers themselves – to Newton
and Faraday, to Poincaré and Einstein, to Darwin and Pasteur, to
Freud and Jung, and Durkheim and Piaget, to rediscover the significance of their contributions to the world they have helped create.

Index

abstract journals 138
accelerators, particle 61–4, 66
action
  concordant with belief 105–18, 119
  guided by science 2, 32, 140, 158, 183
  sensorimotor 112, 114–18, 156, 180
  social 180, 181, 184
aesthetics 49, 86, 87
algebra 13, 17, 19, 82, 163–5
Allen, L. 72, 74–5
analogy 22–5, 156
anatomy 46, 50, 90, 103, 161
anomalies 34–6, 39, 140–1, 147
'anomalous water' 72–4, 139, 142
anthropology
  as science 158–9, 169, 176, 178, 184
  origins of science in 4, 121, 179
anthropomorphism 156, 181
apes 96, 105, 111
apparatus 57–64, 126
Arabic 12, 117
architecture 99
archive, scientific 17, 105, 106
  access to 88, 137, 140
  contents of 26, 43–4, 50, 85, 102
  incoherence of 22, 38, 91, 100, 138
  quality of 126, 137–8
  re-assessment of 130, 143
  selection for 70, 149, 151
  social sciences 158, 184
Aristotle 3
art 49, 86, 108, 118
astrology 148
astronomy
  discovery in 71, 170
  observation in 43, 52, 110, 155
  relevance of 107
  theoretical discrepancies in 36, 67–8
  theoretical models in 14, 23, 24
  theoretical speculation in 37, 142
astrophysics 28–9
atom
  as category of nature 150
  as chemical symbol 164
  as hypothetical entity 168–9

  as physical object 29, 40, 84
  as planetary model 23, 33
  as thema 156
atomic physics 83, 89
authority
  biblical 134
  conflict of 139
  of scientific community 126
  of scientific knowledge 130
  of scientific theories 91
  of scientists 31, 95, 125, 140, 141, 158
average IQ 165
axiomatization 15, 21–3, 163–4, 173
Azande 120, 134, 148
Aztec 110

Bacon, F. 3
bacteriology 57, 88
Barnes, B. 119, 120
Beethoven, L. van 71
behaviour, animal 85
behavioural sciences
  categories in 28
  commonsense in 184
  credibility of 10, 121, 158–9
  empathy in 177, 185
  experiment in 167
  ignorance in 148
  mathematics in 14, 164, 174
  models in 41, 169
  statistical uncertainty in 69–70
  theoretical incoherence of 38
Behavioural Sciences, Encyclopaedia of 2
behaviourism 115, 122
belief
  and action 105–7, 119, 159
  erroneous 8, 108, 140–1
  from intuition 104–5
  justification of 133
  in natural order 121
  parascientific 144
  in science 5, 109, 121–3, 137–41
  scientific 95, 108, 138–9, 141
  sociology of 88
  in teachers 129, 138

Bell, J. 71
Bellow, S. 185
Bernstein, B. 119
big science 61
biochemistry 50, 157
biography 185
biology 10, 110, 121, 159
  categories in 30, 122, 135, 166
  cellular 50, 161
  classification in 43, 46, 55, 161
  evolutionary theory in 23
  experiment in 57, 127
  molecular 52, 80, 84, 135, 137, 161
'Black Cloud' 27, 122
Blake, W. 42
Blanc, M. 142
blindness 104, 114
Bloor, D. 13, 16, 107, 120
Blondlot 142
Bohr, N. 23, 24, 33, 40, 172
books 106, 127–8, 133, 146
bootstrap principle 157
Boyle's Law 64, 172
botany 29, 44, 46, 50
brain, human
  compared with computer 99, 149
  mathematical models of 22
  neurophysiology of 96–100, 111, 113,
    179
  as scientific instrument 161
Braithwaite, R. B. 100
branching pattern 155
breakthroughs 89, 132
Brookhaven National Laboratory 65
Brown, R. 112, 113, 115
Bruner, J. S. 104
Bulmer, R. 161
Bunge, M. 101, 102, 104

caloric 25, 93
Campbell, D. T. 52, 131, 137
Canetti, E. 168
Capaldi, N. 4
Capra, F. 85, 157
Cassirer, E. 82, 178
categorial framework 16, 21, 32, 121, 135,
    138
categories
  behavioural 162, 171, 174, 183
  consensible 166
  of everyday reality 135
  formal, sharp 26, 160, 162
  of natural language 113, 121
  of nature 33, 150, 156, 161
  paradigmatic 132, 183
  physical 28, 181
  social 161–2, 164, 172, 180
causality 84, 181
cell 29, 50, 150, 161

cerebration 149
CERN 65
chemistry 10, 28, 29, 90, 110
  atomic hypothesis 168
  atomic models 51–2, 79
  experimental technique 127
  neutrino detection 36
  periodic table 31
  physical 73
  symbolism 85, 163–4
chess 174
child development
  cognition 104–5
  language 111–15, 136, 156
  material reality 124, 127, 129, 137
  sensorimotor coordination 114, 156
  social reality 180–1
chimpanzees 111, 112, 140, 176
China 13, 46, 110
Chomsky, N. 19, 113, 116
citations 130, 132
classification 43, 46, 160, 161, 181
Clausewitz, K. von v, 174
coelacanth 71
cognition, human 6, 95, 109, 179
  cultural variation of 116–19, 121
  individual 8, 87, 101, 129
  individual variation of 119, 122, 151
  mathematical theories of 22
  transformational 105
Cole, J. R. 130
Cole, M. 117, 119
Cole, S. 130
Coleman, J. S. 166, 169, 173
Collingwood, R. G. 177
Collins, H. M. 68
colour perception 117, 120
commonsense
  evidence 9, 91, 92
  logic 116
  realism 120–1, 147–8
  science as 124–5, 129, 135
  social 178, 184
communication, human 118, 119, 129, 137,
    181
communication, scientific 4, 42, 100, 123,
    159, 184
  categorial framework 21, 90
  data selected for 61
  language of 6, 11, 26, 111
  mathematics in 15, 28, 30
  noetic domain of 106
  official system of 132–3, 143, 145
  pictures in 85, 111
  theories in 7
community, scientific
  controversy in 133
  orthodox Establishment 135, 142–4
  paradigms shared by 7, 87–8, 108–9

community, scientific (*cont.*)
  sociology of 3, 4, 7, 110, 126
  validation by 42, 59, 70, 72, 125, 131, 148, 159, 184
competition, scientific 59, 63, 133
complementarity 136, 145
computer
  chess playing 174–5
  data processing 66, 113
  intelligence simulating 98, 100
  interactive 54–5, 99
  memory 78, 80, 106
  pattern recognition 53–4
  powers of 20, 22, 149–50, 171
  'superhuman' 33
concepts 89, 118, 151
  reality of 127, 185
  thematic origins of 156–8
  transformation of 102
  of velocity 96
Condillac, E. 27
confirmation
  of behavioural theories 169, 172
  of discovery 139
  of prediction 7, 10, 20, 31, 58, 163
  of theory by experiment 33, 54, 56, 63–4
  as unsuccessful falsification 38, 71
conjectures 22, 40, 82, 93, 98, 122, 184
  status of 37, 148
  validation of 30, 131, 140, 159, 171
conjuring 121, 147
consciousness 32, 86, 106, 156–7
  of society 178–80
  and speech 112, 114–15, 120
consensibility 6, 27, 99, 104–5, 145
  empathic 178, 181–3
  of mathematics 13, 31, 38, 173
  of observation 26, 42, 59–60, 77, 166
  of perception 7, 30, 43, 46, 55, 94
  of psychic phenomena 147–8
  of social categories 162, 171
consensuality 6, 90, 105, 110, 147, 151
  of archive 137–8
  of behavioural sciences 159, 162, 171, 182, 184
  of classification 46
  of language 11
  of mathematics 13, 19, 30, 160, 173
  of observation 42, 59, 77, 166
  of perception 54, 78, 86, 87, 177
  of themata 156
  of time order 92
  testing for 7, 56, 108
consensus
  current 127, 129, 138
  everyday 120–1
  goal of science 3, 6, 7, 30–1, 41, 56, 108, 124–6, 166, 176
  as paradigm 88–92

  theoretical 77
consensus gentium 179
conservation laws 156
  of energy 145, 165
  of mass 110, 164
Constable, J. 86
context 152, 154
continental drift 93, 133–5, 139, 152
contradiction 37, 40, 93, 108, 116, 131, 139, 141
controversy 35, 40, 57, 59, 67, 132–3, 139, 143–5
correspondence principle 136
cosmology 37, 68, 91, 107, 116, 131
cost-benefit analysis 165
cranks 145
creativity 5, 20, 59, 88, 102, 131
credibility of science 100, 109, 121, 139–40, 143, 158
credulity 91, 121, 148
criticism 7, 21, 40, 59, 88, 108–9, 126, 131–2, 134, 142–4, 171, 185
culture 109
  diversity of 111, 116–19, 161, 179, 182
  language and 112, 113
  primitive 131, 134
  traditional 119

'Daleks' 122
Dalton, J. 164, 168–9
Darwin, C. 23, 46, 157, 185
data processing 66
deaf-mutes 104, 114
de Gaulle, C. 12
de Grazia, A. 143
Denbigh, K. G. 92
Dennett, D. C. 100
Derjaguin, B. 73
Descartes, R. 3
diagrams 85–7, 96, 104
Dilthey, W. 177
Dirac, P. A. M. 25
disciplines 8, 129, 134
discovery 7, 58, 70–6, 147–8, 170
  erroneous 73–5, 131, 142
  psychology of 88, 129
  teaching method 71, 126
  unconfirmed 67–8, 72, 139
disposition property 169, 170
dissent 131
DNA 84, 121, 125, 161, 169
dogmatism 92
dolphins 96, 105, 111, 176
doubt 5, 57, 59, 108–9, 122, 129, 140
Downs, R. M. 104
Durkheim, E. 181, 183, 185

Eccles, J. C. 181
economics 3, 158, 159, 165, 171, 173–4, 182

Eddington, A. S. 28
Edison, T. 142
editor 145
education, scientific 88–91, 119, 126–30,
    135, 138, 183
  discovery method 71
  historical method 83
Einstein, A. 5, 21, 43, 67, 103, 145, 146,
    153, 157, 185
electromagnetism 25, 29, 128, 153
electrons 25, 80
  as category of nature 29, 150
  reality of 63, 124–5
elephants 96
emotion 2, 10, 86, 159, 176, 178–81, 185
empathy 119, 177–81, 185
empirical
  knowledge 76, 109
  relations 163
  statements 27, 31, 39
engineering 2, 5, 10, 28, 61, 99, 145, 184
English 11, 12
enigmas 148, 154
epidemiology 70
epistemology 2, 14, 61, 63, 64, 100, 102
  evolutionary 131
error, scientific 7, 93, 108–9, 120, 123, 139
  correction of 41, 132
  experimental 35, 64, 66, 75
  human 87, 141
Eskimo 117
ether 23, 25
ethology 85
European science 110, 131
Evans-Pritchard, E. E. 134
evolution
  theory of 23, 25, 46, 110, 140, 155, 157,
    161
  human 180
  model of science 131, 136, 140
experience
  everyday 91, 122, 124
  logic of 26–8
  personal 86, 90, 106, 114, 116, 127, 128,
    183, 185
  research 40, 62, 90–1, 102, 127, 185
  social 163
experiment 6, 7, 10, 37, 56–60, 137
  bias in 139
  crucial 58
  fit with theory 33–5, 39
  psychological 168, 176
  repetition of 132
  simplification by 166
  social and behavioural 70, 166–8, 176
  techniques of 26, 31, 127
  unsuccessful 130
expertise 91, 108, 125, 130, 134, 138, 143,
    184

extrapolation 107
extra-sensory perception 146–8
eye, human 96, 99, 116, 117, 161

factor analysis 170
faith 37, 129
fallacy 8, 93, 120, 126, 129, 132, 139–40
fallibility 39, 92, 108, 140
falsification of theories 8, 9, 10, 35–8, 40,
    57–9, 61, 64, 69, 71, 147, 167, 170, 173
fashion 139
Ferguson, E. S. 45
Fermi surface 80
Flammarion, C. 142
fluoridation 138
formalism 163–4, 172
four-colour theorem 150
fraud 140–1
French 11, 12
Fresnel, A. J. 142
Freud, S. 185
Friedlander, M. W. 146
fundamentalism 134
future 32, 33, 38, 92, 106, 148

Galen 46
Galileo 11, 41, 110, 142, 143
Gallagher, C. F. 12
Galvani, L. 142
games 105, 173–6
Gardner, B. T. 112
Gardner, R. A. 112
gases, theory of 24, 64, 84, 145, 165, 172
Geertz, C. 179, 180, 182
Geller, U. 147
Gell-Mann, M. 20, 31, 54
Gellner, E. 131, 134, 136
genetics 9, 25, 110, 124, 135, 161, 169, 170,
    182
genius 146, 147
geography 82–5, 155, 175
geology 10, 28, 29, 52, 71, 78, 90, 93, 122,
    130, 133, 134, 135
geometry 13, 14, 19, 20, 80, 96, 116, 127,
    128, 157, 163
George, W. H. 78, 145
Gillette, R. 146
'Go' 174, 175
God 30, 62, 100, 101, 110, 149
Gödel's theorem 14, 27, 139
Gombrich, E. H. 91, 98, 118, 151
Goodenough, W. 182
Goody, J. 11
Gould, S. J. 144
grammar 15, 19, 103, 112–16, 122
graph theory 156
gravitation 67–8, 72, 157, 170
Greece 13, 15, 169
Greeley, R. 55

Greenfield, P. 118
Gregory, R. L. ix, 98, 137
Gurwitsch 142

Hallam, A. 92, 133, 152
Hamilton, W. R. 21
Handy, R. 173
Hanlon, J. 147
Hanson, N. R. 45, 55, 102, 137, 152
Harmon, L. D. 96
Harré, H. R. 103, 183, 184
Hartmann, W. K. 144
Hasted, J. B. *et al.* 147
heat 24
Heisenberg, W. 24, 40, 66
Helmholtz, H. von 124
Hempel, C. G. 14
Heraclitus 27
heresy 133
Heritage, J. 177, 184
Hertz, H. 90
Hesse, H. 105
Hesse, M. 25, 40, 78
heuristic model 140
heuristic principle 175
Hewish, A. 71
hidden variables 39, 168–71, 172, 184
Hildebrand, J. 73
history 178, 185
 of science 22, 57, 83, 88, 93, 126, 130,
  136, 151, 157, 163, 169
Holton, G. 103, 154, 156
Hopi 117
Horton, R. 121
Hoyle, F. 122
human nature 180
humanism 177–85
humanities 2, 185
Hunt, E. B. 98
Huxley, T. H. 185
hypothesis 6, 7, 22, 30, 38, 83, 131, 139,
  148, 151, 168
 false 92–3
 inadequately tested 140
hypothetical entities 72

icon 103, 104, 183
ideology 136, 182
illusion, optical 153
image processing 66, 115
imagination
 scientific 5, 24, 32, 36, 91, 132, 134, 185
 sources of 88, 101, 151
 perceptual 104
Inbar, M. 175
India 13, 85, 162
indoctrination 90, 127–9, 134–5
induction 32; 99, 127
infant mortality 69, 170

information, scientific 78, 82, 88, 96, 130,
  132, 137, 149
 false 140
 visual 96
 as commodity 165
Inhelder, B. 104, 111, 114
insight 2, 168, 175, 185
inspiration 88, 147
instruments
 humans as 96, 120
 objectivity of 57, 123
 observational 42, 50, 53, 67–8, 170
 scientific 35, 56, 60–4, 77, 91, 177
intelligence 96, 98, 104
 artificial 98–9, 100
 extraterrestrial 110
 measurement of 164–5, 170
interpolation 82
intersubjectivity 7, 8, 100, 119, 181
 categorial framework for 21
 humanistic 177–9
 of language 15, 30, 102, 104
 of observation 46
 of pattern recognition 43, 55
 of themata 156
introjection 105, 180
introspection 101, 115–16, 122, 177
intuition 20, 23, 101–5, 135, 150, 156–7,
  185

Japan 11, 97, 119
Jarvie, I. C. 180, 183
jellyfish 122, 161
Joule, J. P. 142
journalism, science 143
journal, scientific 138, 144, 145
Jung, C. G. 95, 104, 185

Kac, M. 19
Kant, I. 3, 15, 30, 78, 156
Kapitza, P. L. 31
Keith, A. 140, 141
Kelvin, Lord 23
Kendall, D. G. 82
Kennedy, J. 182
kinship relations 155, 159, 180
Klein, F. 82
knowledge
 behavioural 158, 183
 chemical 80
 consensual 120, 158, 163
 empathic 178
 empirical 76
 ineffable 103
 objective 6, 8, 50, 107
 organized 185
 personal 6, 11, 86, 90
 practical 159
 psychological 121

knowledge (*cont.*)
  public 17, 49, 70, 88, 102, 106, 111, 147
  qualitative 85
  reliable 77, 84, 183
  scientific 6, 30, 45, 64, 78, 83, 99, 105, 149
  social 121, 158, 167, 177, 180, 184
  superior 106
  tacit 23
  theoretical 87
knowledge, sociology of 64, 119, 120, 134, 182
Knox, R. 101
Koestler, A. 143, 144
Körner, S. 14, 16, 26–8, 121
Kuchowicz, R. B. 35
Kuhn, T. S. 40, 58, 78, 83, 89, 136, 185

Lagrange, J. L. 21
Langmuir, I. 142
language
  categorial 89
  comprehension of 104
  consensibility of 119, 179
  faculty for 114
  foreign 12, 112, 127
  formal 111
  learning 112
  logic of 114, 136
  mathematics as 13–14, 18, 30, 102
  natural 12, 14, 104, 106, 111–16, 118, 120–2, 124, 135
  neurophysiology of 97
  of science 6, 27, 127, 145
  sign 112
  as social medium 128
  spoken 118
  unambiguous 130
  visual 79, 96
  written 118
Laplace, P. S. 33, 149
Latin 12
law 2, 12–13, 32
laws of nature 21, 26, 29, 30, 32, 83, 84, 107, 110, 130, 182, 185
Lavoisier, A. 142
layman 91, 125–6, 128, 137, 139, 141, 143, 145, 158
learning 104, 112, 127
Leatherdale, W. H. 25
Lee, T. D. 58, 71, 151
LeGuin, U. 179
Leonardo da Vinci 46, 49
libraries 106
life, origins of 91
linguistics 4, 19, 102, 111–16, 118
liquids 24, 37, 59, 73, 172
literature, scientific 138
Logan, J. L. 67

logic
  of communication 100, 103
  of everyday world 122, 147, 160
  formal 19, 78, 120, 162
  in Japan 119
  of language 12, 113, 115, 120
  of mathematics 15–16, 82, 94, 150, 160, 163, 172
  of perception 88, 177
  of science 7, 43, 99, 137, 160
  of sensorimotor action 114, 127
  of the situation 130, 174
  of social experience 164, 178, 180, 181
logic, two-valued and three-valued 26–8, 30, 35, 64, 100, 122, 160, 162–3, 171–4
Lorentz, H. A. 25
Luria, A. R. 97, 113, 180
Lyons, J. 19, 113
Lysenko, T. D. 133

Macaulay, T. B. 158
McCusker, C. B. A. 73
McNeill, D. 102
magic 120, 121, 134, 146, 148
Maglich, B. 65
Malthus, T. 157
Mannheim, K. 185
map
  abstract 85, 164
  behavioural 169
  brain 115
  games 175
  geological 79
  imaginary 108
  material 77–82, 111
  mental 114
  as metaphor 78, 82–5, 108
  multiple-connectivity of 82, 84
  paradigm as 89, 117
  becomes picture 89, 90, 102, 103, 116, 128
  scale of 84
  scientific 140, 149, 151, 159, 183
  sketch 85, 104, 171
  of social domain 180, 182
  weather 149
  world 20, 77–94, 100, 107
map making 132, 160, 184
Mars 54, 149
Marshall, J. C. 102
Marx, K. 185
material domain 106, 111, 114, 116, 119, 120, 122, 137, 156, 173, 180
materials science 28, 52–3
mathematics 6, 13–21, 148–50, 172
  applied 18, 21, 55, 163
  diagrams in 54, 82, 96
  Chinese 13
  graph theory 156

mathematics (*cont.*)
  Greek 13
  Indian 13
  intuition in 101
  as language 13–14, 28, 38, 78, 122, 125, 136, 160
  learning 127
  as logic 15–16, 26, 102
  philosophy of 14, 27
  in physics 28–30, 77, 92, 94, 163
  pure 18, 21, 102, 146
  in social sciences 14, 163–5, 172
  as transformation machine 17–20
  verbal definition in 111
Mauss, M. 181
maxim 38, 185
Maxwell, J. C. 25
May, R. M. 150
Mayer, J. 145
Mead, G. H. 26, 63, 128
Mead, M. 177
measurement 16, 27, 33, 46, 58, 85
mechanics, classical 5, 21, 22, 39, 82, 96, 110, 122, 136, 156, 182
Medawar, P. B. 28, 141
medicine 2, 46, 88, 110, 127, 145, 184
Medvedev, Zh. A. 133
memory 22, 32, 97, 98
Menard, H. W. 130
Mendel, G. 169
Mendel'eef, D. I. 31
mental domain 86, 106, 109, 111, 112, 117, 118, 129, 137, 151, 176, 180
mental processes 101, 102, 104, 106, 116, 137, 149
Merton, R. K. 4, 185
meson, A_2 65
metallurgy 28, 80
metaphor 21, 23–5, 27, 78, 117, 151, 156
metaphysics 32, 56, 59, 100, 120
  of science 136, 159
metascience 5
meteorology 22, 149
Michelson, A. A. 33
Michelson–Morley experiment 25, 83
microbiology 50
microscope 50, 52
Minkowski, H. 21
Minnaert, M. 147
miracle 121
Mises, R. von 14
model
  economic 171
  hypothetical 169, 172
  Jerusalem 92
  mathematical 22, 29–30, 163, 172
  molecular 50
  non-deterministic 175
  as paradigm 90

  of reality 71
  realizable 23, 175
  of social behaviour 171, 175, 184–5
  social, of science 3, 6
  as thema 156
  theoretical 22–6, 33, 36, 77, 159
model-building 98, 132, 160
molecule 14, 24, 29, 40, 50–2, 80, 135, 150
Monod, J. 80
monographs 132, 138
monopole, magnetic 72
Morgenstern, O. 173
motion 96, 115, 120
Murstein, B. I. 185
mysticism 157

Naess, A. 36, 109, 137
Nagashima, N. 119
Nagel, E. 14, 178, 184
natural history 43, 46, 50
natural sciences 9, 10, 30, 124, 127, 159–60, 169, 180–1, 183
Needham, J. 47, 110
Neeman, Y. 20, 31, 54
network metaphor 39–40, 83, 88, 140
neurophysiology 43, 78, 103, 122
  of emotion 179
  of language 112–13
  of perception 96–9, 101, 115
neurosurgery 97, 179
neutrinos, solar 35, 139
New Guinea 117, 161
Newman, J. R. 14, 146
Newton, I. 108, 160, 185
Nicod, J. 35, 127
noetic domain 82, 106, 128, 141, 149, 174
  cultural variability 119–21, 182
  language from 111–13
  connected with mental domain 106, 111, 180
  paradigms in 117
  segments of 108
  themata in 157
'noise' 64–71, 139, 170
normal science 58, 83, 89, 90, 132
norms of scientific community 4, 59, 94, 123, 129, 131–2
novelists 185
novelty, scientific 57–9, 88, 129, 131, 136, 142, 151
nuclear tests 69, 170
nucleon 29, 150
nucleus, atomic 24, 29, 37, 60, 83, 125, 150, 172
number 13, 46, 114, 163–5

objectivity 107–9
  cultural relativism of 110
  of everyday reality 122, 135, 147

objectivity (*cont.*)
  experimental 57–60
  instrumental 53, 56, 176
  intersubjective 7, 8, 15, 59, 107–9, 119,
    124, 181
  of logic 16, 181
  of maps 87
  of observation 42
  of personal experience 181
  superhuman 100
observation 6, 10, 42–76, 137
  of behaviour 85, 163
  consensible 159, 167, 176
  discovery by 70–2
  instrumental 60–4, 77, 151
  logic of 26–8
  objective 100, 156
  reproducible 42, 56, 57, 131, 132, 167
  technique of 127
  theoretical fit to 34
  unconfirmed 68, 72, 139
  visual 43, 56
observer
  bias of 168
  empathic 176–8
  equivalent 43, 56, 117, 125, 176, 181
  independent 59, 63, 68, 86, 104, 135
  individual 86, 90, 105
  scientific 76, 88, 95
  social 176–8
  trained 160
Oedipus 182
Oldfield, R. C. 102
Olver, R. 118
Omega-minus particle 20, 31, 54, 71
organelle 50, 150, 161
organism 29, 150, 161
orthodoxy 133, 136, 142

Page, T. 144, 147
pain 181
painting 86, 118
palaeontology 140–1, 160–1
paper, scientific 132, 138–9
Parable 182
paradigm 7, 8, 50, 58, 62, 89–91, 117
  breaking through 132, 151
  commitment to 134
  consensual 90
  education in 126
  formation of 90, 151
  internalization of 90, 105, 128, 183
  of naive realism 136
  physics as scientific 38
  preservation of 132, 136
  replacement 93, 152
  science as cultural 183
  switch 152
parapsychology 146

parascience 143–8
parascientism 148, 183
parity, conservation of 58, 71, 151
Parsons, T. 185
particle physics 20, 25, 53, 58, 65, 72, 91,
  135, 147, 151, 156, 157
Pasteur, L. 57, 88
pathology 50, 161
pattern recognition
  in biology 43–50, 160–1
  by computer 98–9, 150, 177
  consensibility of 7, 42–3, 77, 95, 122, 137,
    151
  context of 155
  in geology 50, 94
  intuition and 103–4
  neurophysiology of 95–7, 115–16
  non-logicality of 16, 100, 122
  in physical sciences 51–6
penicillin 80
perception
  auditory 42, 95
  computer simulation of 98, 177
  consensibility of 42, 49, 179, 181
  cultural variation of 111, 117, 121
  extra-logicality of 99–101
  extra-sensory 146–8
  intuitive 103
  neurophysiology of 22, 96–9, 111, 115
  non-human 91, 122
  primary 106, 112
  psychology of 137, 151
  tactile 42, 95
periodic table 31
pharmacology 179
phenomena, abnormal 146
phenomenological theory 22, 39, 91, 140,
  150
philosophy 2, 3, 16, 53, 85, 87, 109, 125,
  127, 136, 150
  Eastern 157
  of language 102
  of life 147
  of mathematics 14, 27
  of scepticism 109
  of science 3, 4, 22, 32, 57, 64, 78, 84,
    95, 124
phlogiston 93
photograph 50, 51, 55, 56, 63, 77, 118, 154
  bubble chamber 53, 65, 87, 99
  cloud chamber 72
  LANDSAT 87
  as map 85, 87
physicalism 30, 46
physical sciences 18, 28, 33, 77, 121, 159,
  160, 163, 166
physics
  classical 23, 92, 108
  discovery in 58, 72, 170

physics (*cont.*)
  education in 90, 126
  high energy 31–2, 53, 61, 62, 131
  instrumentation 60–4
  laws of 107, 110
  mathematical 20–1, 25, 94, 102, 145, 173
  as mathematical science 18, 28–30, 33, 92, 160, 163
  models in 22–6, 36–8, 172–3, 175
  as paradigm science 9–10, 41, 160, 185
  prediction in 30–4, 58
  quantal 92
  theoretical 3, 22, 40, 45, 172, 175
  validation of 38–40
  visual perception in 52, 80
physiology 46, 50, 90, 96, 110, 111, 117, 161, 179
Piaget, J. 104, 111, 112, 114, 116, 127, 156, 185
picture 7, 8, 27, 45, 54, 85–7, 90, 122
  computer generated 99
  as map 85, 108, 128
  mathematical 102
  paradigm as 91
  scientific world 92, 119, 137, 159
  social world 169, 171, 180
'Piltdown Man' 140–1
Pirsig, R. M. 124
Planck, M. 136
plate tectonics 133
Plato 3
poetry 11, 32, 119, 122, 159, 185
Poincaré, H. 185
Polanyi, M. 23, 24, 36, 43, 78, 102, 103, 106, 109, 127, 134–5, 185
politics
  and science 108, 157, 178, 184, 185
  as science 32, 162, 168, 171, 175, 182
  in science 133, 145
  science *in* 2, 143
'polywater' 73–5
Pope, A. 77
Popper, K. 6, 35, 59, 82, 86, 106, 107, 109, 125, 135, 178, 185
positivism 28, 43, 99, 101, 109
precognition 146
prediction
  confirmed 10, 39, 71, 98–9, 131
  experimental testing of 56, 58, 88
  falsified 8, 10, 38
  mathematical 18, 20, 163–4
  practical 33, 107, 122, 149, 184
  rhetorical power of 7, 30–3
  sociological 169, 172
  theoretical 83–4, 159
  unconfirmed 72, 139, 140
prejudice 144
probability 31, 36, 38, 65, 157

problem-solving 58, 88, 98, 102, 104, 111, 114
progress, scientific 88, 89, 129, 130–4, 143, 150
Proust, M. 185
psychiatry 4, 162
psychic domain 178, 179
psychical research 142, 146
psychoanalysis 104
psychology
  of child development 104, 111, 121, 128–9
  of cognition 78, 122
  experimental 167, 175
  of invention 5, 30, 70
  of love 185
  parapsychological 146
  of perception 43, 95, 99
  of 'reality' 73
  of refereeing 132
  as science 159, 164, 169, 174, 176, 178, 184
  of situation 177
  of small groups 167, 175
  social 4, 158, 165
psychopathology 91, 168, 179
'psycho-physics' 160
pulsars 71, 170
Puthoff, H. 147
Putnam, H. 17
puzzles 58, 173
Pythagoras 142
  theorem of 13

quantification 166, 174
quantum theory 9, 21, 34, 40, 83, 89, 90, 121, 136, 151, 157
quarks 72
quasars 155

radioactivity 69
Radnitzky, G. 5
Ramanujan, S. 146
Randall, J. L. 146
Ranke, L. von 178
rationality 32, 61, 125, 135, 136, 148, 150, 178, 179, 183
reality 119–23, 124
  commonsense 120–1
  of concepts 63, 89, 127, 169
  cultural variation of 118–19
  everyday 9, 87, 122, 135–7, 147, 159
  models of 71
  of models 73, 92
  personal 90, 111, 127
  of scientific map 109, 128, 151, 156
  social 129, 180–3
'recognition' (as scientist) 144, 148
reductionism 150, 165, 168

Index

redundancy of information 82
referee 132, 145
refutation 88
Reich, L. 118
Reines, F. 36
relativism, cultural 113, 120–1, 134, 182–3
relativity theory 5, 21, 25, 43, 67, 83, 88, 121, 136
religion 2, 121, 125, 134, 148, 182
Renaissance 131
reproducibility, experimental 60, 63, 64, 68, 75
research 3, 127, 185
review articles 130, 132, 138
revolution, scientific 22, 72, 89, 91, 93, 132, 133, 151, 169
rhetoric 7, 12, 20, 22, 31, 39, 107, 184
Rhine, J. B. 146
robot 100, 107, 122, 175, 177
Roszak, T. 1
Royal Society 57, 60, 124, 144
Rudwick, M. J. S. 79
Russian 11
Rutherford, E. 23, 31, 33, 60
Ryle, G. 12

Sagan, C. 144, 147
Santayana, G. 9
savage mind 121
scepticism 36, 37, 41, 72, 73, 88, 90, 109, 120, 122, 127, 129, 148, 170, 171, 183, 184
Schon, D. A. 25
schooling 118–19
Schutz, A. 129, 178
science
　applications of 10, 32, 107, 127
　as commonsense 124
　as an Establishment 142
　ethos of 131, 151
　goals of 132
　ideology of 136
　as an industry 60
　lay view of 125, 128
　limitations of 109, 149, 182–5
　map metaphor for 87
　metaphysic of 136
　pathological 4, 133, 142
　as minimum variance culture 183
　as social institution 106, 125, 126
　social model of 3–8, 9, 17, 59, 63, 94, 105, 125, 131, 133, 158, 178
　social role of 1, 3, 105, 126, 133, 141, 143
　theoretical nature of 87
　in totalitarian country 133, 145
　uniqueness of 110
　universality of 109–11
　Western 110, 157, 169
science fiction 179

science policy 143
scientific attitude 109, 124
scientific method 2, 6, 9, 17, 68, 74, 84, 130, 137, 160, 166, 173
scientism 30, 183
scientist
　as author 12, 64, 132
　as authority 31, 125, 130, 139, 158
　as *bricoleur* 157
　as communicator 42, 50, 70, 95, 99, 119
　credulity of 90, 95, 148
　as critic 7, 12, 59, 95, 132
　as 'Everyman' 134
　as experimenter 6, 39, 56, 58, 62, 75, 127, 167
　as expert 95, 125, 130, 131, 132, 138
　fallibility of 141
　as guardian of truth 133, 136
　insanity of 144
　internalizes paradigm 88, 89, 91, 95
　learns to think scientifically 111, 128
　as member of society 105
　as observer 6, 8, 42, 50, 53, 56, 95, 99, 101, 125, 127
　as philosopher 124
　as professional 90, 125, 135, 138, 144, 146
　as referee 132
　as sceptic 12, 36, 59
　as team researcher 60
　as theorist 6, 8, 36, 39, 173
Scribner, S. 117, 119
sectarianism 8, 9, 134
sensorimotor
　action 112, 114–16, 118, 156, 180
　coordination 111, 128, 135, 137
　experience 120, 181
signal detection 66, 70, 71, 140, 170
simulation, social 174–6
Sinclair-de-Zwart, H. 114
Smith, W. 79
social domain 161
social sciences 2, 10, 14, 28, 30, 69, 157, 158–85
society
　open 131
　technical 118, 183
sociology 70, 107, 158, 159, 164, 169, 174, 176, 178, 182, 184
　of belief 88
　of knowledge 64, 119, 120, 134, 182
　of language 111
　of science 3, 4, 60, 64, 75, 126, 185
'socio-physics' 160
space 21, 29, 83, 96, 115, 117, 120, 127, 145, 181
specialization 124–6, 134
speech 78, 112, 118
'spontaneous generation' 57, 93, 135
Sprat, T. 57

statistical analysis 65, 67, 69, 70, 139, 146, 159, 169, 170
Stea, D. 104
stereochemistry 52, 80
Sternglass, E. J. 69, 170
Stoll, C. S. 175
strategy 174
Strawson, P. F. 15
subjectivity 52, 86, 87
superconductivity 40, 135
Suppes, P. 13
surgery 97, 103
surveying 82, 108, 151
Swedish Royal Academy 138
symbolism 17, 19, 85, 163–4
symmetry 20, 25, 58, 71, 156
sympathy 178
synesthesia, verbal–visual 117
systems analysis 171

Takeuchi, H. *et al.* 93
Targ, R. 147
Tart, C. T. 157
taxonomy 46, 55, 86, 161, 162
Taylor, J. G. 147
team research 63
technology 1, 10, 45, 57, 110
telepathy 124, 146
text-books 39, 40, 90, 138, 156
themata 156–7, 169, 181, 183
theorem 13, 17, 19, 80, 82, 102, 150, 173
theoretician 18
theorizing 36, 57, 77, 130, 132, 160, 169, 173
theory 7, 18, 27, 54, 68, 120, 127, 156
  of apparatus 61
  of behaviour 159, 168
  coherence of 100
  of games 173
  as map 77, 150
  social 171
  speculative 91
  synthetic 173
  'toy' 173
thermodynamics 22, 25, 84, 165
Thom, R. 18, 27
time 29, 32, 83, 96, 115, 117, 120, 145, 181
  arrow of 92
topology 78, 82, 85, 115, 116, 163, 175
Toulmin, S. 11, 78, 83
'toy' 172–4
tradition 108, 159
translation 11, 12
tree 155, 156
Trimble, V. 36
truth 8, 14, 26, 28, 66, 70, 100, 105, 109, 120, 133, 137, 140, 146, 152, 158
Turchin, V. E. 119
Turing machine 100, 177

UFOs 68, 144, 147
uncertainty
  experimental 66, 75, 82
  scientific 109, 122, 123, 140, 160, 171
  Heisenberg principle of 66
uniqueness of science 109, 110, 113, 122
universality
  cognitive 117, 120, 121, 129
  of science 109–11, 126, 134

value, human 178, 182–3
Van Leer Jerusalem Foundation ix
Velikovsky, I. 143, 144, 146
velocity 96
Venn diagram 103, 115
verification 59, 75
Vesalius, A. 49
von Neumann, J. 173
Vygotsky, L. S. 112, 180

war games 174–5
Waterston, J. J. 145
Watkins, J. 178
Weber, J. 67–8, 72, 170
Weber, M. 15
Wegener, A. 93–4, 133, 152
Weizenbaum, J. 98
White, P. 185
Whitehead, A. N. 1, 142
Whorf, B. 117
Wigner, E. P. 29, 30
Wittgenstein, L. 3, 15
world
  of conscious experiences 86, 90
  everyday 9, 26, 29, 55, 115, 120, 124, 129, 135
  external 42, 61, 100, 106, 156
  material 104–6, 112, 118
  of nature 89, 95
  physical 28
  real 26, 42, 136
  social 179, 182
world 1 106
world 2 86
world 3 6, 82, 174
world map 20, 89
world picture 7, 8, 32, 90, 119
  scientific 125, 135–8, 157
  non-scientific 120, 125
writing 11, 97, 118

Yamadori, A. 97
Yang, C. N. 58, 71, 151
Young, R. M. 157
Young, T. 142

Zen 122
Ziman, J. M. 1, 2, 12, 34, 131, 141, 172
zoology 177